U0162524

读客文化

| 中英双语版 |

万物简介
黑洞是什么

[英] 凯瑟琳·布伦德尔 著

张建东 译

BLACK HOLES:
A VERY SHORT INTRODUCTION

浙江科学技术出版社

著作合同登记号 图字：11-2022-034

图书在版编目（CIP）数据

万物简介. 黑洞是什么 : 汉文、英文 / (英) 凯瑟琳・布伦德尔 (Katherine Blundell) 著 ; 张建东译. —— 杭州 : 浙江科学技术出版社, 2022.11
书名原文: Black Holes: A Very Short Introduction
ISBN 978-7-5341-9973-8

Ⅰ. ①万… Ⅱ. ①凯… ②张… Ⅲ. ①黑洞 - 普及读物 - 汉、英 Ⅳ. ①P15-49

中国版本图书馆CIP数据核字 (2022) 第051161号

书　　名 万物简介：黑洞是什么
著　　者 ［英］凯瑟琳・布伦德尔
译　　者 张建东

出　　版 浙江科学技术出版社
地　　址 杭州市体育场路347号
邮政编码 310006
发　　行 读客文化股份有限公司
网　　址 www.zkpress.com
联系电话 0571-85176593
印　　刷 三河市龙大印装有限公司

开　　本 880mm × 1230mm 1/32
印　　张 8.25
字　　数 132 000
版　　次 2022年11月第1版
印　　次 2022年11月第1次印刷
书　　号 ISBN 978-7-5341-9973-8
定　　价 49.00元

责任编辑 卢晓梅　**责任校对** 张　宁
责任美编 金　晖　**责任印务** 叶文炀

献给蒂姆和路易斯·桑德斯，

我深爱着你们。

致谢

菲利普·欧科克、罗素·欧科克、史蒂文·巴尔拜斯、罗杰·布兰德福特、史蒂芬·布伦德尔、史蒂芬·贾斯姆、汤姆·兰开斯特、拉莎·梅农、约翰·米勒和保罗·托德为本书的草稿提供了很多有用的意见。史蒂芬·布伦德尔为本书制作了图表。斯蒂文·李协助进行了光学观测。我对他们表示衷心的感谢。

凯瑟琳·布伦德尔

牛津大学

2015 年 4 月

目录

01

黑洞是什么

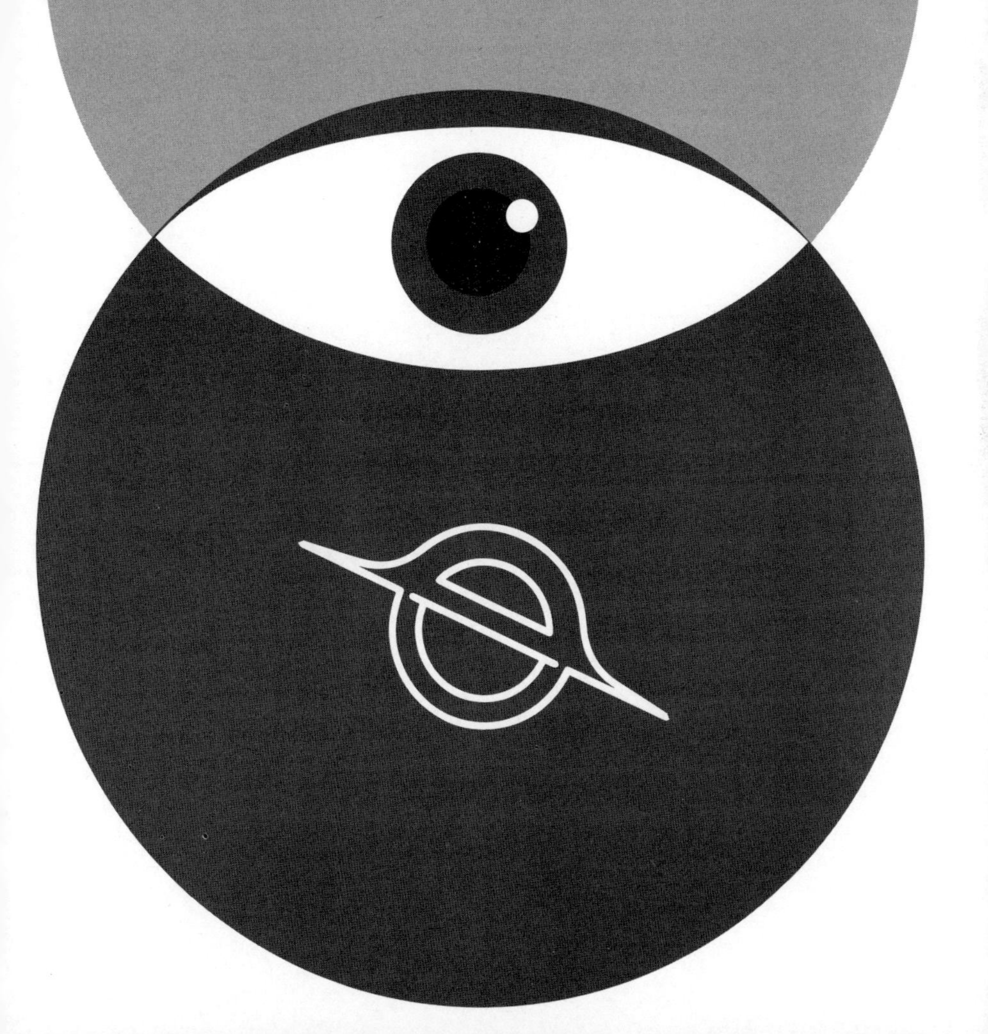

英文页码 1

黑洞就是一个引力很强的空间区域。任何东西——甚至连光都因为不够快，不能从其内部逃离。虽然这一概念最初是由理论物理学家通过丰富的想象构思出来的，但现在我们已经在宇宙中发现了数百个黑洞，还能计算出上百万个。尽管这些黑洞是不可见的，但它们以一种很容易被探测到的方式与周围环境相互作用，并对其产生影响。确切地说，这种相互作用的性质取决于相对黑洞的距离：太近的话是不能逃脱的，但较远的地方就会出现一些极其壮观的现象。

1964 年，安·尤因（Ann Ewing）在一篇报道 1963 年于得克萨斯州举办的一个研讨会的文章中首次提到了“黑洞”一词，然而她从未说明是谁发明了这个词。1967 年，美国物理学家约翰·惠勒（John Wheeler）需要一个词作为“引力坍缩彻底的恒星”的简写，于是开始推广这个术语——不过“坍缩的恒星”这一概念早在 1939 年就由他美国的同事罗伯特·奥本海默（Robert Oppenheimer）和哈特兰·斯奈德（Hartland Snyder）提出来了。事实上，关于现代黑洞概念的数学基础在 1915 年就已经诞生了。

德国物理学家卡尔·史瓦西（Karl Schwarzschild）在空间中物体孤立无转动的质量的条件下解出了爱因斯坦的重要方程（在他的广义相对论中被称为场方程）。

2

在此之后过了20年，印度物理学家苏布拉马尼扬·钱德拉塞卡（Subrahmanyan Chandrasekhar）研究了恒星死亡时会发生什么。以此为基础，英国的亚瑟·爱丁顿爵士（Sir Arthur Eddington）解决了一些相关的数学问题——比奥本海默和斯奈德的工作稍早一点。爱丁顿的计算表明，当大质量恒星耗尽所有燃料时会坍缩形成黑洞，不过爱丁顿自己在1935年向英国皇家天文学会宣称其物理含义是“荒谬的”。尽管这个概念看起来荒谬，但黑洞无疑是我们的银河系乃至整个宇宙物理现实的重要组成部分。1958年，美国的大卫·芬克尔斯坦（David Finkelstein）取得了更进一步的进展，他明确了黑洞周围存在一个单向表面。这对于我们将在下一章中讨论的内容具有重要的意义。这个表面的存在不允许光从黑洞内部强大的引力中脱离，而这也是黑洞是黑色的原因。要理解这种现象是如何产生的，我们首先要理解物理世界的一个深刻特性：任何运动的粒子或物体都存在一个最大速度。

快是有多快

丛林法则之一是：不想死得快，就得跑得快。除非你异常狡猾或者善于伪装，否则只有足够敏捷才能存活下来。哺乳动物逃离险境的最大速度取决于其质量、肌肉力量和新陈代谢之间复杂的生化关系。宇宙中运动最快的实体所能达到的最大速度是由完全没有质量的粒子所呈现的，例如光的粒子（被称为**光子**）。这个最大速度被精确地定为每秒 299 792 458 米，几乎比空气中的声速快 100 万倍。如果能以光速旅行，我将能够在十四分之一秒内从我在英国的家到达澳大利亚，这就是一瞬间的事情。从离我们最近的恒星，也就是太阳出发的光只需要 8 分钟就可以到达我们这里。而从太阳系最外层的行星海王星出发，光子到地球也只需几个小时。我们说太阳离地球有 8 光分，而海王星离我们有几光时。这会导致一个有趣的后果，如果太阳停止发光或海王星突然变成紫色，地球上的任何人发现这些重要信息都分别需要花上 8 分钟和几小时。

3

现在让我们来考虑光线从太空中更加遥远的地方传回地球的时间有多长。我们的太阳系所在的银河系是一个直径达几十万光年的星系。这意味着光从银河系的一侧行进到另一侧需要几十万年。离本星系群（银河系是其中的重要成员）最近的星系团，也就是天炉座星系团，远在几亿光年之外[1]。因此，如果在围绕天炉座星系团中某颗恒星运行的行星上有一位观察者，手头配备了恰当的仪器回看地球，可能会看到恐龙在地球上徘徊。不过这只是由于宇宙浩瀚得令人难以置信，才使得光的运动看起来迟缓且费时。但当我们开始考虑如何将火箭发射到太空时，光速是上限这一“强制规定”就会带来一种有趣的效应。

逃逸速度

如果我们发射火箭时速度太慢，那么火箭将没有足够的**动能**来挣脱地球的引力场。反之，如果火箭的速度恰好足以逃离地球引力的拉扯，我们就说它已达到了**逃逸速度**。火箭从诸如行星之类的大质量物体上逃离时，行星质量越大，火箭距行星的**质心**越近，逃逸速度也就越大。逃逸速度 v_{esc} 可以写成 $v_{\mathrm{esc}}=\sqrt{2GM/R}$，其中 M 是行星的质量，R 是火

4

1 最新研究显示天炉座星系团离地球约 7500 万光年。

箭与行星质心的距离，而 G 是被称为牛顿引力常数的自然常数。重力作用总是将火箭拉向行星或恒星的中心，朝向被称为质心的点。不过，逃逸速度的取值与火箭的质量完全无关。因此，不论其内部载荷是几根羽毛还是几台三角钢琴，从距离地球质心约 6400 千米的卡纳维拉尔角发射的火箭都具有相同的逃逸速度，也就是 11 千米 / 秒多一点或约为声速的 34 倍（可以写为 34 马赫）。现在假设我们可以压缩地球的全部质量，使它占据更小的体积，假定它的半径变为其当前的四分之一。如果火箭发射处距离质心 6400 千米，其逃逸速度将保持不变。然而，如果将它重新放到距质心 1600 千米的压缩后的地球的新表面，那么逃逸速度将会是原始值的两倍。

现在假设某些灾难的发生导致地球的全部质量都收缩到了一个点，我们把这样的物体称为**奇点**。它现在已经成了一个“质点”，一个占据空间体积为零的有质量物体。在距这个奇点只有 1 米左右的地方，逃逸速度将远大于在 1600 千米处的取值（约为光速的 10%）。离奇点更近，略小于 1 厘米的地方，逃逸速度将等于光速。在这个距离上，光本身没有足够的速度来逃离引力的拉扯。这是理解黑洞性质的关键思想。

对“奇点”一词的用法值得深究。我们不相信在

5 持续的引力坍缩的终点，物质会变成什么几何点；正相反，我们会发现经典引力理论失效并进入量子体系。从这里开始，我们将使用术语奇点来指代这种极其致密的状态。

事件视界

现在想象你是一名驾驶宇宙飞船的宇航员，并且正在接近这个奇点。当距离它还有一段距离时，你可以随时将发动机反转并逃之夭夭。但是距离越近，就越难体面地撤离。最终你会到达一个无论装载的发动机有多强大都无法逃脱的距离。这是因为你已经到达了**事件视界**，这是一个用数学方式来定义的球面，它也被定义为内部逃逸速度超过光速的边界。对于我们关于地球坍缩到一个点的思想实验而言，这个表面将是一个以奇点为中心，半径只有 1 厘米的球面，这对我们的太空船来说可能很容易避开。然而当黑洞由恒星而不是行星坍缩形成时，事件视界会变得更大。事件视界有一个重要的物理效应：如果你在那个表面之上或者里面的话，物理定律根本不允许你逃离，因为这样做你需要打破普适的速度限制。事件视界是一个强制性的分界面：在它之外你有决定你命运的自由，而在它之内你的未来将被锁在里面，不可改变。

这个球面半径被称为史瓦西半径，是为了纪念前面提到的卡尔·史瓦西。作为第一次世界大战中的一名士兵，史瓦西得到了广义相对论中著名的爱因斯坦场方程的第一个精确解。史瓦西半径写为 $R_s = 2GM/c^2$，其中 M 是黑洞的质量，G 是牛顿引力常数，c 是光速。根据这个公式，地球的史瓦西半径还不到 1 厘米。以此类推，太阳的史瓦西半径为 3 千米，这意味着如果我们的太阳被压缩成奇点，那么距这一点仅 3 千米之处的逃逸速度就将等于光速。一个质量是太阳质量 10 亿倍的黑洞，其史瓦西半径也就是太阳的史瓦西半径扩大 10 亿倍，因为一个无旋转的理想物体的史瓦西半径与其质量成正比。正如我将在第 6 章中所描述的那样，这些巨大的黑洞被认为存在于很多星系的中心。

在牛顿物理学中，这种对事件视界的描述是合理的。事实上，在爱因斯坦等人出生前几个世纪，类似黑洞的物理实体就被想象出来了。它们深刻地改变了我们对空间和时间的理解。最早想象出类似黑洞的“**暗星**”的人是 18 世纪的约翰·米歇尔（John Michell）和皮埃尔·西蒙·拉普拉斯（Pierre Simon

Laplace），而现在我将解释他们做了什么。

天文学的一个非凡之处在于，即使你被困在地球上，也能发现很多关于宇宙的事情。例如，没有人曾经到访过太阳，然而在19世纪后期，人们通过分析太阳光谱发现太阳中存在氦。需要特别注意的是，这也是氦元素第一次被发现，它在太阳上被发现的时间要比在地球上被探测到早得多。在更早的18世纪，关于黑洞的一些设想就开始形成了，比如关于所谓暗星的概念。人在很大程度上是他那个时代的产物，那些饱含奇思妙想、迈出第一步的人都是这样。

约翰·米歇尔

英格兰的乔治王时代是一个相对和平的时期。英国内战早已过去，英格兰已经成为一个内部相对安宁的国度（距拿破仑在法国的崛起还有一段时间）。约翰·米歇尔（图1）和他的父亲一样，作为牧师接受了大学教育并加入了英格兰教会。作为西约克郡桑希尔的教区长，米歇尔能够继续他的科学研究，把他对地质学、磁学、重力学、光学和天文学的兴趣贯彻下去。与当时在英国工作的其他科学家一样——比如

7

天文学家威廉·赫歇尔（William Herschel）和物理学家亨利·卡文迪许（Henry Cavendish）（米歇尔的密友）——米歇尔能够顺应新的牛顿式思想的潮流。艾萨克·牛顿爵士（Sir Isaac Newton）构想出了引力定律，彻底改变了人们对宇宙的理解。这个定律解释了

图 1　约翰·米歇尔，博学家[1]

---

1 图上文字为：约翰·米歇尔牧师，英国皇家学会院士（1724—1793），地质学家和天文学家，1767—1793 年任桑希尔教区长，主要研究磁学和天文学的实验，设计了扭秤称量世界实验。他的宾客包括亨利·卡文迪许、威廉·赫歇尔、约瑟夫·普雷斯特里（Joseph Priestley）和约翰·斯米顿（John Smeaton）。

太阳系中行星的运转与他那颗树上掉落的苹果，是受到了相同的力的作用。

8

牛顿的思想使人们可以用数学来研究宇宙，而新一代的科学家们也能够将这种新颖的世界观运用到不同的领域。米歇尔特别关注的是利用牛顿的思想，通过测量恒星发出的光来估计它们到附近恒星的距离。为此，他想出了各种方案，比如将恒星的亮度与其颜色联系起来；此外，他还考虑了**双星**（一对处于对方引力束缚中的恒星）以及它们的轨道运动所代表的具体动力学信息。米歇尔研究了恒星在天空中特定区域的聚集情况，他将此与随机分布进行对比，检验并推断出聚集的原因是引力成团。在当时，这些想法没有一个是切实可行的：人们知道的双星很少（虽然赫歇尔正在编制令人印象深刻的各种双星和新天体的目录），而且恒星的亮度和颜色之间的关系也并非尽如米歇尔所认为的那样。不过米歇尔尽力对更广阔的宇宙做了牛顿对太阳系所做的事情：对观测结果进行科学、合理和动态的分析，以此提供关于天体的性质、质量和距离的新信息。

米歇尔一个洞若观火的观点来自这样一个思想实验：用他的话说，就是光粒子是“与我们熟悉的所有物体一样会受引力作用；也就是说，受到与它们的惯性（他的意思是质量）成正比的力。据我们所知，目前还没有任何理由怀疑或相信，万有引力是一个普适的自然定律”。他推断，这些由大恒星发出的光粒子会受恒星引力的吸引而减速。因此，到达地球的星光会变慢。牛顿已经证明了光在玻璃中会减速，这就解释了折射的原理。米歇尔推论出如果星光确实也会减速，那么用棱镜来探测星光就可能会观测到这种减速效应。实验是由皇家天文学家牧师内维尔·马斯基林（Nevil Maskelyne）博士而不是米歇尔做的，他希望观察到星光折射能力的减弱。卡文迪许写信告诉米歇尔这并没什么用，而且“几乎不太可能找到光线显著减弱的恒星”。米歇尔感到沮丧，但是这种天文学推断很难说，极大程度上靠对不可估量的事情的猜测：星光会受到发射出它的恒星的引力影响吗？米歇尔无法确定。但他大胆地作出了一个有趣的预测。

9

如果一颗恒星的质量足够大，并且引力确实影响了星光，那么引力就足以完全控制住光粒子并防止它们逃逸。这样的物体将是一颗**暗星**。这样，这个在约克郡的教区中写作的鲜为人知的牧师就成了第一个构想出黑洞的人。然而在那时，米歇尔测量恒星距离的方法还不完善。更重要的是，他此前一直对自己的健康漠不关心，这令他无法继续使用望远镜。卡文迪许给他写了一封信安慰他："如果你的健康状况不允许你继续使用（望远镜），我希望它至少可以让你更轻松省力地权衡（称量）这个世界。"卡文迪许这唯一的笑话（他以沉默寡言而闻名）指的是米歇尔构思的另一个思想实验——"称量世界"实验，即让扭秤两端的大铅球被两个固定的铅球所吸引，从而用测量引力强度的方法，推断出地球的重量。以前从来没有人这样做过。米歇尔的想法非常棒，但他生前并没有完成这个实验。米歇尔的实验后来由卡文迪许代劳，现在被称为卡文迪许实验。这一荣誉被转嫁给卡文迪许，但卡文迪许也丢失过很多荣誉，没有发表自己许多具有突破性的研究，被归功于后来的研究者（包括"欧姆定律"和"库仑定律"的提出）。

皮埃尔·西蒙·拉普拉斯

10　在英吉利海峡的另一边，皮埃尔·西蒙·拉普拉斯没享受到英国启蒙运动和平时期的宁静田园风光。拉普拉斯经历了法国大革命，不过他的职业生涯因协助成立新的法兰西学院和综合理工大学而蓬勃发展。有一段时间，他甚至担任过拿破仑统治下的内政部长，这是一次短暂的任命，因为皇帝后悔了。拿破仑意识到拉普拉斯是一流的数学家，但作为管理者还达不到平均水平。拿破仑后来在谈到拉普拉斯时写道："他到处寻求微妙之处，只考虑问题本身，最终把'无穷小'的精神带进了政府"。拿破仑有其他管理人员可以任命，但世界上几乎没有几个数学家像拉普拉斯那样高产且富有洞察力。他在几何学、概率论、数学、天体力学、天文学和物理学方面都作出了重要贡献。他研究的主题多种多样，包括毛细作用、彗星、归纳法、太阳系的稳定性、声速、微分方程和球谐函数等。他思考过的其中一个领域就是暗星。

1796 年，拉普拉斯出版了他的《宇宙体系论》（*Exposition du système du monde*）。这本书是为受过教育的公众撰写的，书中描述了天文学所依据的物理原理、万有引力定律和行星在太阳系中的运动方式，以及运动和力学定律。这些观点适用于各种现象，包括潮汐和岁差，书中还包含了拉普拉斯对太阳系起源的推测。书中有那么一段与我们的故事密切相关。拉普拉斯计算出类似地球的物体需要有多大，才能使其逃逸速度等于光速。他的计算相当准确：当一个天体密度和地球相当但半径是太阳的 250 倍时，它的表面引力会让光都无法逃离。因此，他推断宇宙中最大的物体是看不见的。它们是否仍然潜伏在黑暗的夜空中无法被探测？还是像我们幻想的那样："外面"只有我们能看到的那些明亮发光体？匈牙利天文学家弗兰茨·萨韦尔·冯·扎克（Franz Xaver von Zach）请求拉普拉斯提供导出这一结论的计算方法，拉普拉斯还帮忙将其（用德语）写出来并发表在冯·扎克担任编辑的一本期刊上。

后来，拉普拉斯慢慢了解到了光的波动说。米歇尔和拉普拉斯的想法都部分基于光的粒子说。如果光由微小的粒子组成，那么这些粒子会受到引力场的影响，并且将永远被束缚在质量大小足够的恒星上，这一结论似乎是合理的。但是在19世纪早期，有许多实验似乎更能证明光的波动说。如果光是一种波，那么就会更加难以观察到引力对它的影响。拉普拉斯对于暗星的预测在《宇宙体系论》后来的版本中被悄然省略了。毕竟米歇尔和拉普拉斯一直在对理论进行推测和探索，而不是致力于解释观测结果，因此这个想法被遗忘了一段时间。米歇尔和拉普拉斯所想象的物体就是"暗星"，这种宇宙中的庞然大物可以凭借其质量维持行星系统，但同样，其夸张的尺寸也使其无法通过光被观测到。从米歇尔和拉普拉斯认为的暗星表面发出的星光太过缓慢，无法克服强大的表面引力。米歇尔和拉普拉斯没有想到的是，这种庞大的质量累积将会不稳定并坍缩。而且在坍缩的过程中，它们会刺穿空间和时间的结构并产生奇点。因此“黑洞”不是“暗星”，接下来的讨论会涉及黑洞的天文发现，但我们首先需要了解时空的本质。

时空

12

我们的日常经验告诉我们：有形宇宙可以通过一个时间坐标 t 和三个空间坐标来界定（例如沿着三个相互垂直的轴 x、y 和 z[1]）。1905 年，爱因斯坦发表了他关于狭义相对论的革命性论文，阐述了运动和静止的相对性。1907 年，赫尔曼·闵可夫斯基（Hermann Minkowski）阐述了如何借助四维时空来更深入地理解这些结果。四维时空中由四维坐标（t、x、y、z）所确定的点对应着“事件”。这个事件是在特定时间（t）和特定地点（x、y、z）所发生的事情。这种被称为闵可夫斯基时空的四维坐标，精确地指明了事件发生的时间和地点。爱因斯坦的狭义相对论可以用闵可夫斯基时空来表述，并为不同参照系中相对运动的物理过

1 这是由勒内·笛卡尔（René Descartes）发明的概念，被称为笛卡尔坐标。

程提供了一种方便的描述方法。“参考系”仅仅是某个特定观察者所拥有的视角。爱因斯坦称这种理论为“狭义”，是因为它只涉及一个特定的情况，也就是无加速的参考系（称为惯性坐标系或参照系）。狭义理论只能应用于匀速运动的非加速参考系。如果你扔下一块石头，它会加速落向地面。在石头上的参考系是一个加速的参考系，因而不能用爱因斯坦的狭义理论来处理。有引力的地方就会有加速度。

这一缺陷促使爱因斯坦提出了广义相对论，该理论在狭义理论提出 10 年后公开发表。他发现，虽然笛卡尔空间和闵可夫斯基时空是物体“生活、移动和存在”于其中的刚性框架，但时空实际上是一个敏感的实体：它可能会因质量的存在而弯曲或变形。

13

一旦质量存在于物理情境中，描述了现实的行为就会不可分割地互相联系起来。这被约翰·惠勒（John Wheeler）简练地总结为：

- 物质作用于时空，告诉它如何弯曲。
- 时空作用于物质，告诉它如何移动。

这种特性是由广义相对论中的爱因斯坦场方程所量化的，该方程将时空曲率与引力场联系了起来。

物理学家认为大质量物体周围存在**引力势阱**。图 2 所示的漫画显示了时空在一对黑洞附近是如何变形的，其中每个区域的弯曲方式都可以被认为是与其质量也就是与引力本身直接相关。时空中的奇点可以被认为是时空中的曲率变得非常高，而使你超越了经典引力体系，进入了量子体系的地方。奇点周围的事件视界起到了单向膜的作用：粒子和光子可以从外部宇宙进入黑洞，但没有任何东西可以从黑洞的视界内逃逸到外部宇宙。事实上，质量并不是黑洞可能拥有以及被测量的唯一属性。如果黑洞在旋转，也就是说它会自旋，那么就会出现更极端的行为。在研究这个问题之前，我们将稍微拓展一下，学习一下如何用示意图来表示时空本身。

14

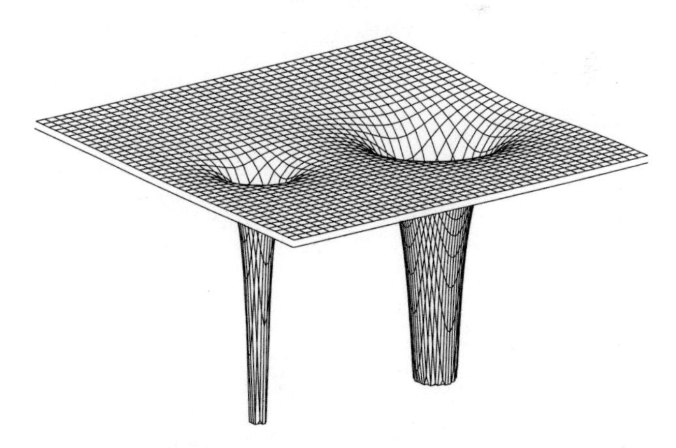

图 2　时空由于物质的存在而产生的变形，也就是时空的弯曲

游览时空

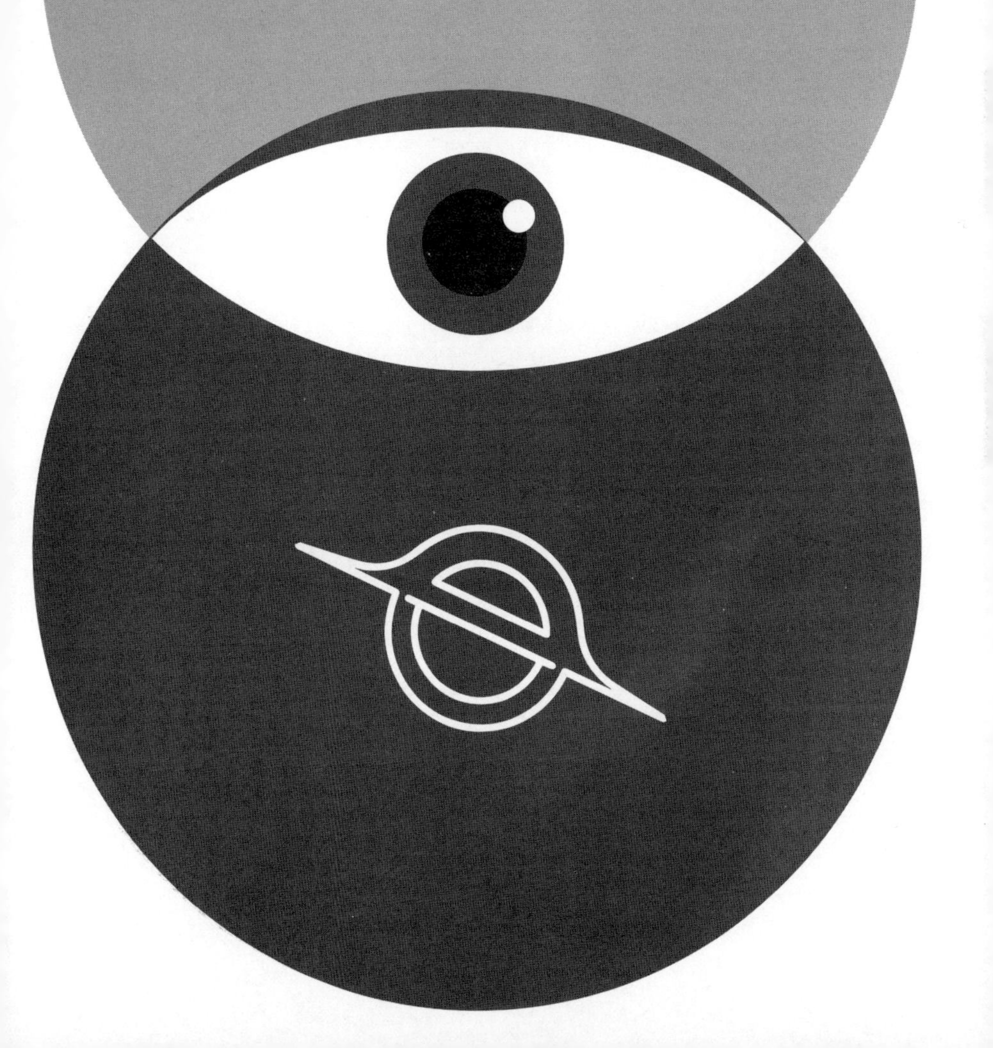

15

数学是一种精致而完美的语言，它可以描述相对论是如何适用于物理宇宙和整个时空的，而这种描述包括了在黑洞附近所发生的奇怪行为。虽然数学的描述强大而准确，但对于那些没有经过适当的专业培训的人来说，也可能是一门令人生畏的外语。描述性的言辞无论多么雄辩，都缺乏数学方程式的严谨和力量，可能不精确，也有局限性。然而图像（据说是）胜过千言万语，它不仅是一种有效的折中方案，也是一种非常有用的将发生的事情可视化的方法。因此，花一些精力来理解这种被称为时空图的图像是非常值得的，这将有助于我们理解黑洞周围时空的性质。

时空图

图 3 展示了一个简单的时空图。依照传统，“类时”轴是在页面中竖直的那一个，而“类空”轴与它垂直相交。当然，我们确实需要四个轴来描述时空，因为有三个类空轴（通常表示为 x、y 和 z）和一个类时轴。

16 不过两个轴就足以实现我们的目的了（而且四个相互垂直的轴也画不出来）。这两个轴相交的地方被称为原点，对于构建时空图的观察者来说，就是他“此时此地”所在的点。一个理想化的瞬时事件，例如按下相机快门，发生在特定的时刻和特定的空间位置。这样的一个瞬时事件由时空图上的一个点表示，对应于所讨论的时间和空间位置。图 3 中有两个点，它们在空间上是分开的（它们不会发生在空间轴上的同一点），但它们是同时出现的（它们在时间轴上具有相同的坐标）。你可以想象，这两个点对应的事件是两个摄影师同时按下快门，他们站在彼此相隔一定距离的地方，拍摄同一个景象。如果点代表事件，那么时空图中的线代表什么？一条线只是显示出一个物体在时空中通过的路径。在我们生活的过程中，我们穿越时空，并在身后留下轨迹（有点像蜗牛在身后留下闪闪发光的黏液痕迹），它是时空中的一条线，用专业术语来说就是**世界线**。如果你整天待在家里，你的世界线就是一条穿越时空的垂直路径（例如，空间坐标 =“金合欢大道 22 17 号”）。你在时间上向前，但是在空间中不动。另一方面，如果你进行了一次长途旅行，你的世界线会因为你的距离随着时间改变而倾斜，因为你在空间和时间上都进行了移动。

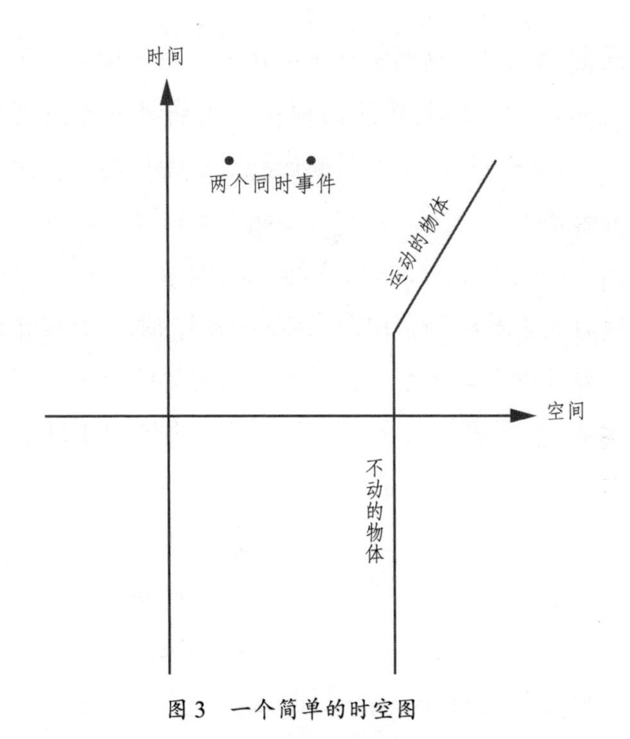

图 3　一个简单的时空图

例如，请看图 3 所示的世界线，这条线的一部分是垂直的，然后再向上变为倾斜。这对应于某个实体的世界线，在垂直的线所对应的时间范围内，这个实体是静止的。举个例子来说，某位摄影师将相机丢在椅子上，由于相机的位置没有变化，所以相机的世界线是垂直的；然后它被偷走了，这时它的空间位置开始连续变化，这条线倾斜的部分就是它的空间位置随时间变化的部分。这条线的斜率会告诉你距离相对时间的变化率，这通常被称为速度。在这种情况下，这是小偷带着赃物逃跑的速度。小偷逃走的速度越快，或者说他在给定的时间内经过的距离越远，倾斜的部分角度就越大。当然，小偷带着赃物逃离的速度有一个固定的上限，就是光速，正如第 1 章所讨论的那样。光束的轨迹将由最大限度倾斜的线（通常通过巧妙设计的单位大小，让最大倾斜世界线在图中与时间轴成 45 度角）来表示。因为没有物体的速度能够超越光速，所以没有世界线可以与时间轴形成更大的夹角。

时空图上的世界线具有一个最大倾斜角度，对应光速这个最大速度，引出了被称为**光锥**的重要概念。这个概念非常简单：你只能通过一些前因对宇宙产生影响，并且因果关系的传播速度不能快于光速。因此，你此刻的“影响范围”被包含在一个有限的时空区域内，即如图 4 所示的与正时间轴成 45 度角以内的部分。此外，你只能受到传播速度不高于光速的事件的因果链的影响。这就是说，只有位于与负时间轴成 45 度角以内的事件才会在当下影响你。如果我们现在用两个类空轴和一个类时轴绘制时空图，那么图 4 中的三角会变成如图 5 所示的圆锥，这就是我们所说的光锥。图 5 中的光锥描绘一个观察者（视作他

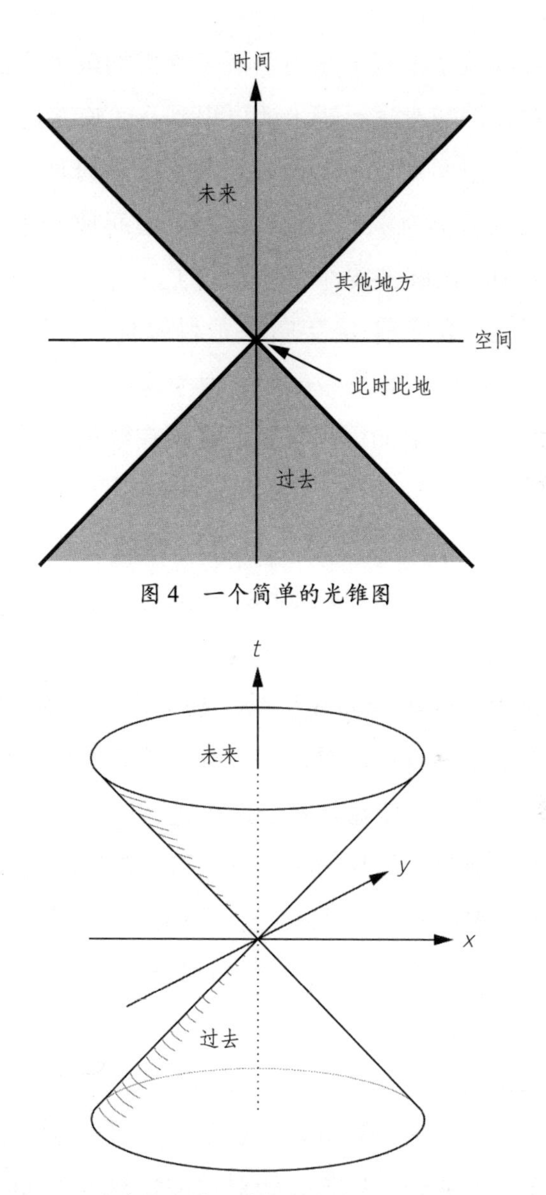

图 4　一个简单的光锥图

图 5　一个表现了某个特定观测者的光锥的时空图

此时此地位于原点）原则上不需要借助破坏宇宙速度的限制，即超光速行进，就可以到达（或在过去已经到达）的空间区域。以正时间轴（未来的时间）为中心的区域被称为未来光锥，而以负时间轴（过去的时间）为中心的圆锥被称为过去光锥。

因此，公元前 44 年暗杀尤利乌斯·恺撒（Julius Caesar）是你过去的一部分，因为这个事件与你之间存在着可以想象的因果联系（必须在学校学习，这证明了因果关系的存在！）。因为来自仙女星系的光可以到达地球上的望远镜，所以它也是你过去的一部分。然而，光需要 600 万年才能到达我们这里，所以 600 万年前的仙女星系是你过去的一部分并坐落在你的光锥上。今天的仙女星系，甚至是公元前 44 年的仙女星系都在你的光锥之外，都不能在此刻影响你，否则因果联系的传播就必须比光速更快。

19

到目前为止，我们在本章中看到的三个时空图，它们的轴被标记为时间和空间。事实上，专业人士通常不会在时空图中画上轴的标签，甚至连轴也不画。时间轴竖直，空间轴水平是大家默认的，但这并不会让专业天体物理学家都草率马虎（虽然这也不是件稀罕事），实际上是因为所有观察者都无法就时空中的确切位置达成共识。在狭义相对论的世界中，同时性的概念被打破了：一个观察者看到两个事件同时发生，并不完全意味着它们对其他观察者来说也是同时发生的。

因此，两个摄影师“同时”按下他们的相机快门，在一个相对相机快速行进的飞船上的观测者看来，这两件事并不是同时发生的。这个观察者推断出的结论将是一个相机在另一个之前被按下。在图3中，我画了垂直高度相同的两个点（之前我断言两个事件同时发生），但出现在快速行进的观测者的时空图上时，它们将出现在高度不同的位置。爱因斯坦的相对论强调说这位宇航员的时空图和我的一样有效。因此，如果时空图上的点位取决于观测者的视角，即他们的参照系，那何必把它们画出来呢？

20

关注一个运动粒子的世界线有助于我们理解这一点；我们现在将绘制一个新的时空图，其中一个粒子带着它的光锥在时空中移动（这个技巧被称为使用共动参照系）。请注意，在图6中，粒子的路径，即其世界线始终保持在光锥内部，因为它的运动速度不能超过光速。

爱因斯坦的狭义相对论是他的广义相对论的一个子集，适用于一组有限的物理情境。在时空不断扩张的前提下，我们需要一个超越狭义相对论的概念框架，一个突出的例子就是不断膨胀的宇宙。在这种情况下，因果关系的表象即为：你无法运动得比你身边局部空间里的光速更快。

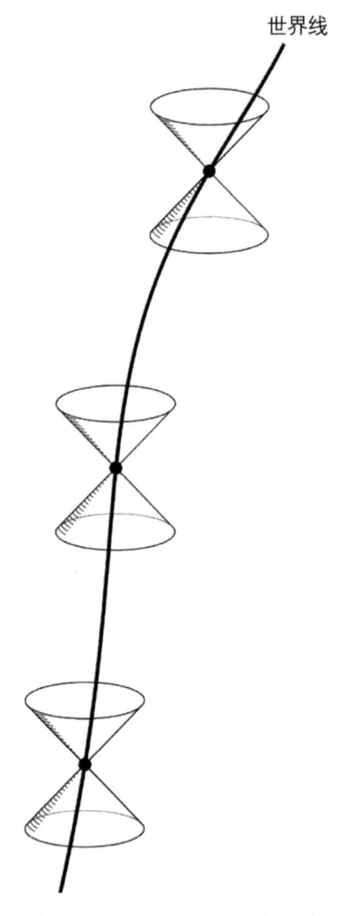

图 6　沿其世界线运动的粒子的时空图。该粒子总会被包含在其未来光锥的内部

物体怎么知道要去哪里

虽然光子没有质量，但事实证明它们仍然受到引力的影响。不过最好不要把这看成某种力的作用，它是由时空的曲率导致的。通常人们认为光子沿着直线行进，这让我们得到了“光线”的概念。然而在弯曲的时空中，光子的行进路径被称为**测地线**。尽管测地线（其名称来自测地学，即测量我们的行星表面陆地的位置）的含义是基于地球的，但它也是描述整个宇宙中时空性质的重要概念。如果空间没有弯曲［情况就完全等同于我们学校教的，欧几里得（Euclid）或他的继承者之一会用的普通几何学］，那么测地线将是光线走过的“直线路径”。但两点之间的最短距离，即光线“想要”走的路线，术语是“零测地线”。在弯曲的空间中，两点之间的最短距离并不是我们所预想的直线，而是测地线。测地线是弯曲空间中的“直线”。直线也可以被描述为“你保持在相同的方向上移动时的路径”。通过比较球面上的经线，可以看出曲面上的几何学有多么不同！如图 7 所示，两条相邻的经线（在赤道处彼此平行）将在极点处相遇于一点。然而在平直空间中，平行线只有在无穷远处才会相遇（依据欧几里得的最后一个公理）。

21

22

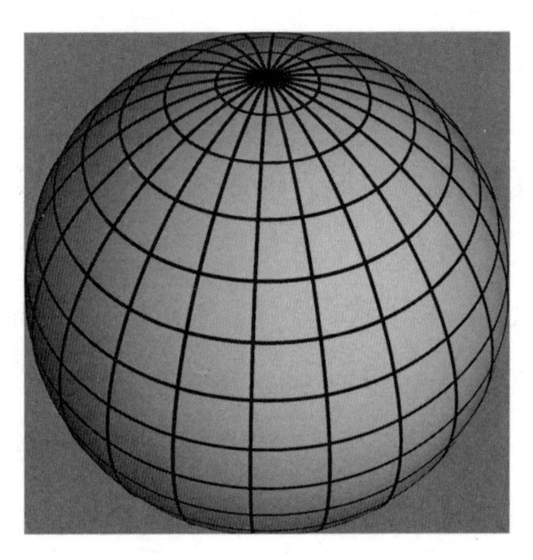

图 7　球体上的经线在赤道处平行，在极点处相交

事实上，在由于质量的存在产生时空弯曲的地方，不受任何外力影响而自由运动的光线或“测试粒子”（物理学家使用的一种假想的装置）在两个事件之间移动时，这种曲率实际上会在移动的路径上体现出来。两个事件应该被视为四维时空中的两个点，每个事件以 x、y、z、t 的形式表示。

23

被称为**度规**的制度规定了我们如何用时钟和尺子测量空间和时间中事件的间隔，它还为解决几何学的问题提供了基础。一个简单的例子是毕达哥拉斯定理，它告诉我们如何计算平面上两个点的距离。爱因斯坦场方程的解则告诉我们如何在物质的分布已知的情况下计算时空的度规。我们用这种方法来构造真实宇宙的测地线。例如，广义相对论的第一批观测证据之一就是在日食期间测量太阳导致的星光弯曲（日食是测量靠近日面的恒星表观位置的好时机，因为来自日面的光被月亮挡住了。1919 年，亚瑟·爱丁顿爵士抓住了一个机会）。太阳的质量会弯曲时空。因此，从遥远的恒星到地球上的望远镜的最短路径（测地线）并不是一条直线——如图 8 所示，它被太阳的引力场弯曲了。

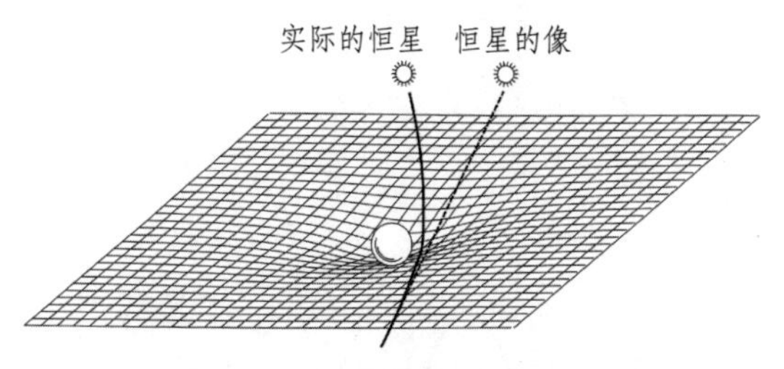

图 8　诸如太阳之类的大质量物体会在时空中引起变形或弯曲

星光的弯曲表明空间是弯曲的，但爱因斯坦的广义相对论告诉我们实际上弯曲的是时空。因此，我们可以想到质量对时间也有一些奇怪的影响。实际上，即使是地球的引力场也足以使地球上的时钟比外太空的时钟运行得慢一点，尽管变化很小但可以测量（大约十亿分之一）。黑洞事件视界附近的引力效应要强得多。因此，即使对于最简单的非自旋黑洞，它附近的时间流逝也与离黑洞很远的时间流逝相差甚远。这

24

是一个真实的属性，并不会随着测量方式不同而变化（例如用原子钟或是电子表）。时间流速的改变直接来自由质量引起的时空曲率，这种效应会使光锥向有质量的物体倾斜。图 9 显示了这种情况的大体效应。

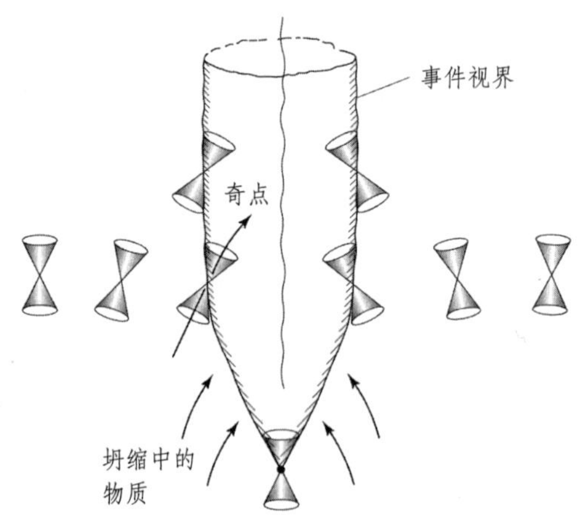

图 9　黑洞周围的时空图。显示了事件视界上物体的未来光锥是怎样位于事件视界内的

黑洞会显著影响光锥的倾斜方向。粒子越接近黑洞，它的未来光锥越向黑洞倾斜，因此黑洞会越来越不可避免地成为其未来的一部分。当粒子穿过事件视界时，其未来所有可能的轨迹都在黑洞内终结。而粒子刚好位于事件视界上时，光锥会大幅倾斜，以至于其一侧与事件视界平行并且其未来完全位于事件视界

25

以内，并且不可能逃出黑洞。图 9 本质上是“局域时空图”的代表，因为这些光锥可以让你知道处于不同位置的测试粒子所经历的局域条件。在这个图中，时间沿着页面向上增加，所以这个图也表现出了黑洞是如何形成并因坠入的物质而增长的。

第 1 章中，我们讨论了米歇尔和拉普拉斯的暗星，它们可能在周围的轨道上维持行星系统，像我们的太阳系一样，实际上我们说的是黑洞。我们只能通过它的引力拉扯而知道附近有一个黑洞。这可能会让你以为表征黑洞的唯一属性就是它的质量。事实上，黑洞是否在旋转会对其性质产生巨大影响，而我将在第 3 章中解释这种影响是如何产生的。

03

黑洞的特征

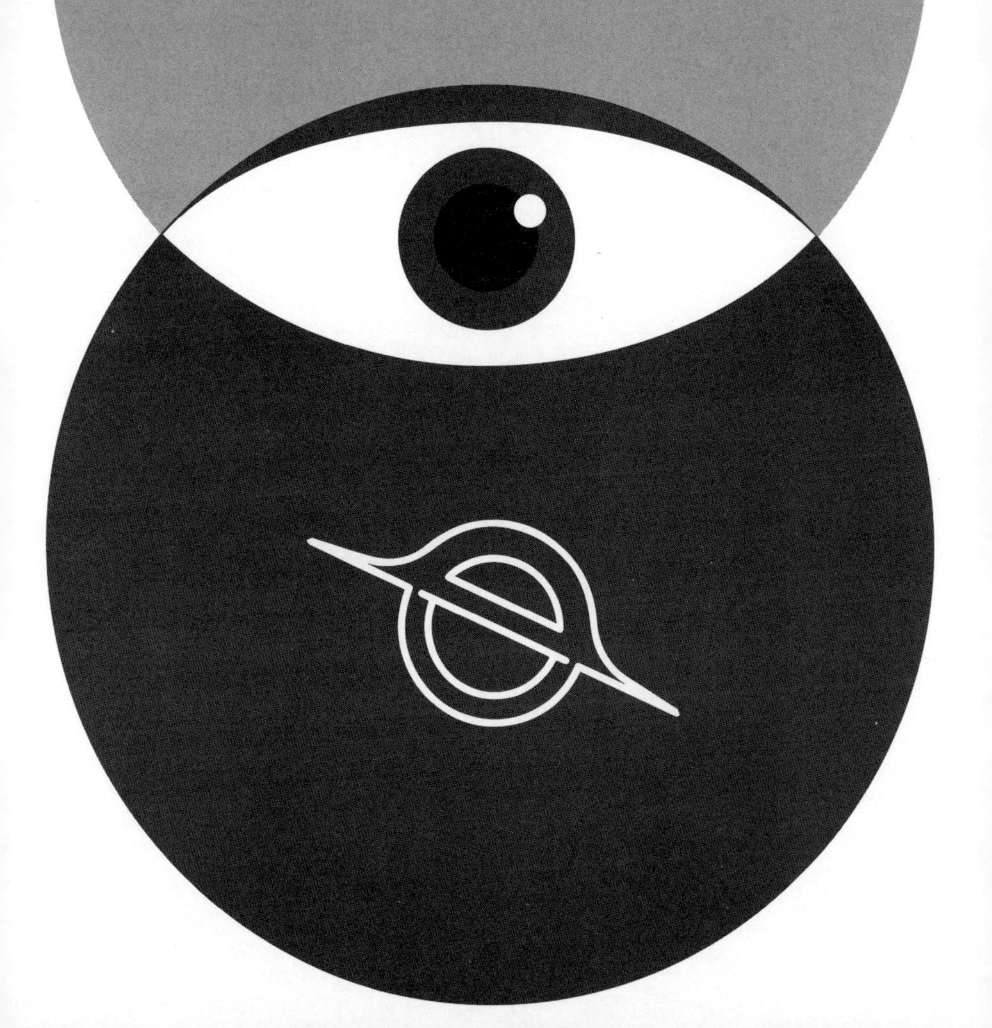

我们在第 1 章中介绍了质量奇点的概念，它由引力坍缩形成，并被一个叫事件视界的界面所包围。这种天体中不自转的被称为**史瓦西黑洞**，这个名词专门表示不转的黑洞：用术语来说就是它们没有**自旋**。简单地说，除了位置以外，能够将两个史瓦西黑洞区分开的唯一特征就是其质量有多大。我们将会在第 7 章了解黑洞是如何生长的，但现在只要知道引力作用下的坍缩是关键因素就够了。如果坍缩前的物质在旋转，那么无论它转得多慢，在发生坍缩时转速都将增加（除非发生某些意外阻止旋转的发生）。这是由一个被称为**角动量守恒**的重要物理定律导致的。这个定律可以通过一个正在做单脚旋转的滑冰者来说明：当她收回手臂时就会转得更快。同样，如果产生黑洞的恒星在缓慢旋转，那么它最终形成的黑洞旋转将会非常显著，这种黑洞被称为**克尔黑洞**。实际上，大多数恒星都在旋转，因为它们原本是由缓慢旋转的大质量气体云引力坍缩形成的。如果最初的气体云有哪怕很小的净旋转，坍缩成的云都将具有非零的角动量，而随着其体积越来

26

27 越小，坍缩成的物体的最终旋转也越来越快。因此我们可以看到，对于新生的黑洞，常有被称为自旋的转动，这即便不是普适的性质也是广泛存在的。我们现在相信，在天体物理中的真实自旋是不可避免的，就像在当今政治中的倾向性描述一样（尽管在后一种情况下，它并非来自角动量守恒[1]）。

我们现在已经阐明了黑洞的第二个物理参数，即自旋或角动量。自旋和质量一样，可以用来区分不同的黑洞。因此，当我们研究黑洞的行为时，这两个性质是很重要的：质量和自旋。理论上，黑洞还可能存在与其行为有关的第三个特性：电荷。这也是物理学中的守恒量，而电荷之间被称为静电力的力与引力有许多相似之处。两者关键的相似之处在于，在大尺度上它们都是平方反比定律的例证。就是说有两个大质量物体，当它们彼此的距离增加到原来的两倍时，它们受到的引力将减小到原始值的四分之一。引力和静电力的关键区别是，虽然引力总是吸力，但静电荷只在有些时候是吸力（当两个物体电荷相反时，即一个带正电而另一个带负电），而在其他时候则是斥力（当两个物体电荷相同时，无论是正电还是负电，

1 原文为“spin is as inevitable in real astrophysical black holes as it is in current-day politics”。此处 spin 一词有“自旋”和“倾向性报道”两个意思。

它们都会互相排斥）。如果两个带电物体带有同种电荷，那么虽然引力倾向于吸引，但静电斥力将倾向于阻止它们聚集。因此，虽然电荷原理上可能是黑洞的第三个属性，人们希望能以之测量黑洞，但实际上黑洞携带的电荷会迅速被周围的物质所中和。因此，一个很好的假设是黑洞只有两个可以明确区分的特性：质量和自旋。就这样！

现在，你可能想知道是否可以通过不同黑洞的成分来区分它们。比如一个黑洞可能是由氢气云形成的，另一个则是由氦气云形成的。为什么坍缩产生黑洞的物质并没有在随后形成的黑洞的可测量性质中体现呢？那是因为信息无法逃出事件视界！光是信息传递的手段，但我们已经在第 1 章中看到光无法从黑洞的事件视界里面逃脱出来。因此从外界看来，落入黑洞的物质的化学成分对黑洞的性质没有影响。将引力看成需要“逃离”黑洞的东西是不正确的。黑洞外部的引力场是随着黑洞形成过程中的时空弯曲而产生的。事件视界形成后，黑洞内部的情况将无法影响外界。

28

黑洞无毛

当我们描述一个人时，经常提及的一个显著特征就是他们的发色（例如金红色、银灰色、巧克力棕色）。有时人们的头发会含有关于他们年龄或国籍的线索。其他有关身体特征的信息如“体重指数”，可能会提供有关其饮食健康的信息。与人类不同，黑洞这个实体除了质量和自旋以外，完全没有其他明显特征（基于前文所述的原因，这里忽略了电荷）。为了强调黑洞没有保留任何关于其前身天体性质的这一特征，约翰·惠勒创造了短语“黑洞无毛”。这里指黑洞不含有原来的形状、原来的团块结构、原来的地形、原来的磁性、原来的化学成分，什么都不包含。白俄罗斯物理学家雅可夫·泽尔多维奇（Yakov Zel’dovich）带领团队进行了计算，结果表明，如果一个表面起伏不平的非球形恒星坍缩成一个黑洞，其事件视界最终会稳定成没有任何团块或起伏的平滑均衡形状。所以，一个黑洞从来不会有不愉快的一天[1]！你唯一可以知道的关于黑洞的事情就是它的质量和自旋。

29

1 此处为 a bad hair day 的双关，hair 可以指黑洞的参数，而 a bad hair day 是指不愉快的一天。

自旋改变现实

引力场会将物体拉到旋转黑洞的旋转轴周围，而不仅仅是朝向它的中心，也许这就是旋转黑洞最引人注目的特征。这种效果被称为**参考系拖曳**。在黑洞的引力场中自由落下时，径向落往克尔黑洞的粒子将获得非径向的运动分量（转动）。

对于具有自旋的测试粒子（例如一个小型陀螺仪）来说，这意味着如果它向着旋转的大质量物体（例如克尔黑洞）自由下落，它的自转轴将会发生变化。就好像中央大质量物体的旋转拖曳了自旋测试粒子的局部参考系一样。这种现象在 1918 年被发现，被称为伦泽–蒂林效应（Lense-Thirring effect），实际上它不仅发生在黑洞的周围，在某种程度上也发生在任何旋转物体的周围。如果你把一个非常精确的陀螺仪放在环绕地球的轨道上，那么参考系拖曳会导致陀螺仪的进动[1]。

爱因斯坦场方程对黑洞进行了数学描述，如第 1 章所述，在稳态（无转动）黑洞的情形下，卡尔·史瓦西解出了这些方程。鉴于史瓦西解出方程是在 1915

---

1 进动是指旋转物体在外力的作用下，自转轴绕某一中心旋转。

年，也就是爱因斯坦引入广义相对论的同一年，可以说这是一项了不起的成就。在很久之后的1965年，新西兰人罗伊·克尔（Roy Kerr）才解决了旋转黑洞的问题。几年之后，澳大利亚人布兰登·卡特（Brandon Carter）进一步探索了克尔的解。卡特深入研究了克尔度规[1]的效应，他证实了由于参考系拖曳，旋转黑洞会在周围的时空中产生巨大的旋涡。旋风属于一种旋涡，靠近旋风中心的空气旋转很快，会带动沿途的任何东西，无论是干草地里的干草还是沙漠中的沙子。在离旋风远一点的地方，空气（以及干草或沙子）会旋转得慢一些。旋转黑洞周围的时空也是如此：远离事件视界的地方，时空旋转的速度很慢，但在视界处，时空的旋转速度与视界自旋的速度相同。

旋转（克尔）黑洞的事件视界与非旋转（史瓦西）黑洞的事件视界非常相似，只是黑洞旋转越快，引力势阱就越深：相同质量的克尔黑洞和史瓦西黑洞，前者形成的引力势阱比后者形成的更深，因此克尔黑洞会是比非旋转黑洞更强大的能量源。我们会在第7章中重新介绍这一点。同时，为了帮助理解这种现象，我们总结出了一个结论：史瓦西黑洞的事件视

1 即克尔得到的旋转黑洞的度规。

界仅取决于质量，但克尔黑洞的事件视界取决于质量和自旋。

有一个特别的问题：即使只在理论层面上，是否存在未被包裹在事件视界中的时空奇点，也就是所谓的“裸奇点”。根据定义，爱因斯坦场方程的所有黑洞解都有事件视界，正如第1章所示，光无法逃离这种视界，因此信息也不行。我们认为所有的黑洞奇点都封闭在事件视界之中，也就是说不是“裸”的，所以宇宙的其余部分无法获得有关奇点的直接信息——这就是英国数学家罗杰·彭罗斯（Roger Penrose）所表述的宇宙监督猜想。他提出所有通过正常的初始条件形成的时空奇点都被事件视界所隐藏，并且在空间中没有裸奇点。 31

太大的自旋有多大

黑洞的角动量大小是有限制的。这个极限取决于黑洞的质量，因此质量大的黑洞比质量小的黑洞旋转得更快。接近它最大极限的旋转黑洞被称为极端克尔黑洞。如果你试图通过加速一个黑洞的旋转来制造一个极端克尔黑洞，可以朝它射入高速旋转的物质（也就是搅拌一下），那么旋转黑洞产生的离心力甚至会阻止物质进入事件视界。

离旋转黑洞的事件视界稍微远一点的地方是另一个重要的数学界面，被称为**稳态极限面**。如果某个大质量物体的自旋不为零，那么在其稳态极限面内就不存在静止的观察者，这叫作对惯性参考系的拖曳：稳态极限面以内的每个可实现的参考系都必须旋转。在这个界面内，空间旋转得非常快，以至于光本身也必须与黑洞一起旋转，而不能保持静止。稳态极限面和事件视界之间的这个区域被称为能层，但很令人困惑的是能层并不是球形的[1]，如图 10 所示。在赤道方向上，能层要比事件视界大得多，但在两极方向上能层与事件视界的半径相同。这导致了能层的形状是扁球形的，类似一个去掉梗的南瓜。能层的前两个音节来自希腊语 érgon，它与“工作”或“能量”有关［比如“人体工程学（ergonomics）”一词］，旧的能量单位尔格（erg）也来源于此。值得注意的是，希腊语中还有一个表示围绕和远离的动词 ergo，也很符合能层的性质。也许这就是在给这个旋转黑洞周围的区域起名并推广开来时，罗杰·彭罗斯和季米特里奥斯·赫里斯托祖卢（Demetrios

1 能层英文为 ergosphere，其中 sphere 的意思是球面。

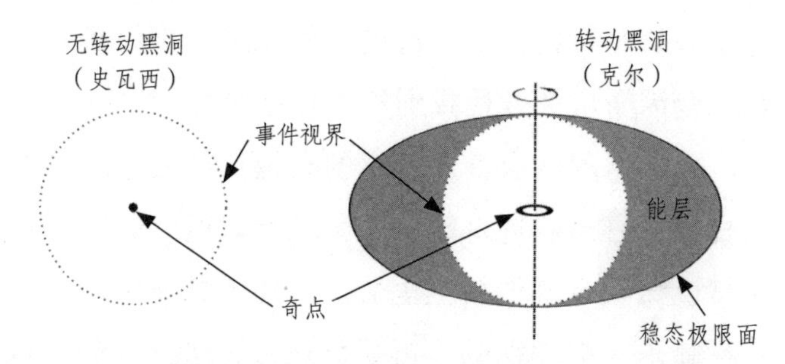

图 10　史瓦西（静止）黑洞和克尔（旋转）黑洞周围不同的界面［在常用的“博耶 – 林德奎斯特（Boyer-Lindquist）”坐标中表示］

Christodoulou）脑海中的依据。能层的重要之处在于在这个区域能够从黑洞中提取能量。

由于能层内的空间在旋转，所以这部分空间内的物质粒子也被迫转动。因此，在这个空间旋转中存储了可观的转动能，这是一件非常重要的事情，我们将在第 8 章中重新讨论它。

白洞和虫洞

广义相对论的爱因斯坦场方程非常丰富，可选择不同的解来描述各种版本的弯曲时空。这产生了几乎取之不尽的平行宇宙，可以提供给宇宙学家描述和思考。我们实际居住的宇宙是哪种类型只能通过观察决定（如果可行的话）。但这并不能阻止理论物理学家利用爱因斯坦场方程找到各种有趣的解。

33 数学物理学家梦寐以求的有趣物体之一就是所谓的**白洞**。白洞表现得就像一个黑洞，但时间是倒流的（可以想象一部倒放的电影）。物质并非被吸入，而是被喷出。事件视界不再是你永远无法逃脱的区域，恰恰相反，它标出了任何东西都无法进入的区域。一旦物质从白洞中出来，它就永远不能再返回那里了——它的整个未来都在外面。正如我们在第 6 章中看到的那样，黑洞是由坍缩的恒星形成的，并且依据量子力学理论最终会形成**霍金辐射**（见第 5 章）。另一方面，白洞只能产生于因某种原因自发聚集成的黑洞的辐射。我们很难理解这在现实中是如何发生的，而且道格拉斯·厄德利（Douglas Eardley）已经证明了白洞本质上是不稳定的。

20 世纪 30 年代，爱因斯坦和他的学生内森·罗森（Nathan Rosen）在研究爱因斯坦场方程时，发现了一个有趣的解。如果一个时空区域弯曲得足够厉害，也就是说它折叠得足够厉害，时空中两个之前被分开很远的部分就可以如图 11 所示，通过一座小桥或者说虫洞连接起来。对于那些希望让人类在宇宙舞台上大展拳脚的作家来说，恒星和星系之间的遥远距离一直是件烦心事，而虫洞（也称为爱因斯坦–罗森桥）为作家提供了一个完美的推动情节的工具，让他们可以把英雄和反派传送到不同的地方。这项数学上的发明对于科幻小说作家来说绝对是一个福音，因为它为穿越太空中遥远的距离提供了一种便捷的手段，从理论上支持了各种高度虚构的和难以置信的飞行器。但回过头来，我们还没有观测到任何表明宇宙中存在虫洞的证据。此外，有相当多的理论证据表明，虫洞形成后不会长时间保持稳定。为了让虫洞持续开放，我们可能需要大量的具有负能量的物质，而所有常规物质都具有正能量（这与引力通常总是吸引这一事实有关）。穿过虫洞的常规物质就足以使虫洞不稳定，然后破坏它，使其变成一个黑洞奇点。

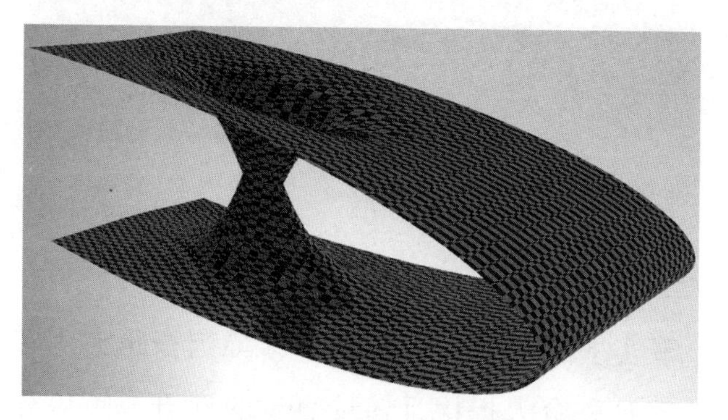

图 11　一个连接两个独立的时空区域的虫洞

如果虫洞确实存在，并且可以维持一段合理的时间长度，那么它们将具有一些令人惊讶且匪夷所思的特性。虫洞不仅可以提供一种在广阔的空间中任意穿梭的捷径，而且可以让旅行者及时返回。于是人们就可以构造在时空中循环的闭合类时曲线，曲线上的光锥会形成一个环（图 12）。这就像在电影《土拨鼠之日》（*Groundhog Day*）中那样，沿着闭合类时曲线行进的人将会简单地一遍又一遍地重复同样的经历。

35

图 12　一个封闭的类时循环，在这个循环中，你的未来将成为你的过去

事实上除了虫洞，还有许多爱因斯坦场方程的解具有这种令人担忧且违反直觉的性质。1949 年，数学家库尔特·哥德尔（Kurt Gödel）发现了一种描

述旋转宇宙的解，其中包含的闭合类时曲线与《土拨鼠之日》的循环完全相同，这些曲线在无尽循环中反复穿过同一个事件（显然“自由意志”不是场方程的一部分）。克尔解中有描述事件视界之外的时空的部分，我们认为这在现实世界中具有真正的物理意义。与此同时，克尔解在数学上关于事件视界内的部分是合理的，但目前还不清楚其是否具有任何物理上的相关性。在克尔解的这一部分中，奇点不是一个点（在非旋转黑洞中是一个点），而是快速转动的环的形状（不过其现实有效性还仅限于推测）。这种环状奇点被闭合类时曲线所包围。在这样的曲线上，你的未来也是你的过去，理论上你有可能在你的父母出生之前谋杀你的祖父母！因此，闭合类时曲线的存在让与时间旅行相关的各种悖论成为可能。对此一种可能的解决方案是承认我们没有将量子力学（描述非常小的物体）和广义相对论（描述非常大的物体）联系起来的理论，也就是量子引力理论。我们不知道极重却极小的物体的物理特性。大多数物理学家认为我们需要这种理论，才能充分理解非常接近奇点的地方的时空特性。因此，或许爱因斯坦场方程的这些奇怪的解并不真正存在于宇宙中，因为它们被基本的量子力学性质所禁止。例如，量子效应可能会使虫洞不稳定。斯蒂

36

芬·霍金（Stephen Hawking）认为情况确实如此，并将这一原则称为“时序保护猜想”。他讽刺道：这是可以保证宇宙免于历史学家侵害的基本原则。

旋转黑洞的内部有许多事情都在我们对基础物理理解的极限之外，因此我们的很多描述都是高度依赖推测的。相比之下，黑洞的自转及其对周围环境的影响，对于我们理解用望远镜所看到的东西具有巨大的现实意义。因此，接下来我们将更加详细地考虑当物质落入黑洞时会发生什么。

04

落入黑洞……

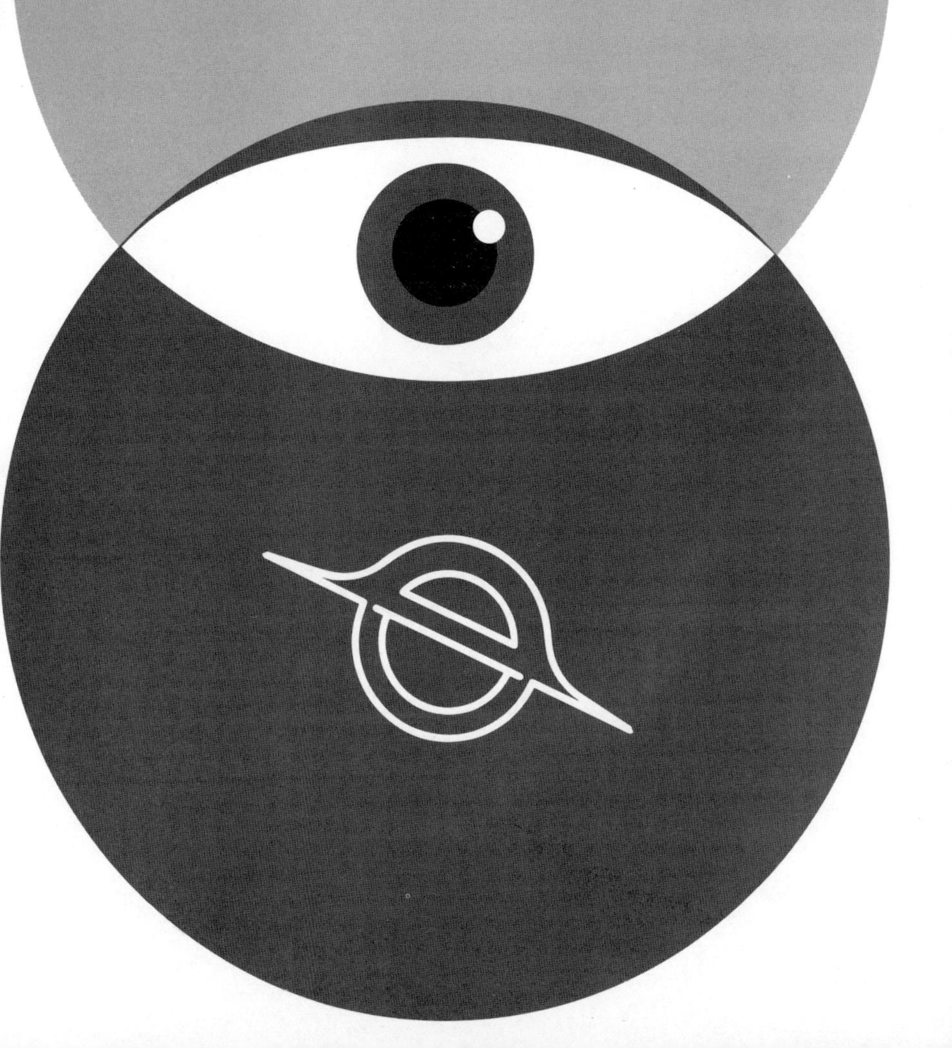

太近是有多近

[37] 如果你或者你的东西不幸落入黑洞会发生什么？在我们详细考虑这个问题之前，了解观测者的特定视角或参考系的影响是很重要的。这意味着不同的观测者看到的东西非常不同。你如何看待落入黑洞的物体完全取决于你离这个物体有多远（以及你是否就是那个物体）。在黑洞的事件视界之外的一个光子，因为它在视界之外，所以理论上可以逃离。而在事件视界内，故事将会不同了——光子无法逃离黑洞的引力场。但即使在事件视界之外，离开黑洞的光子也不会毫发无损地逃脱。由于光子必须克服引力，因此能量会损失。这是引力势阱的一个例子，就像你将自己从深井中拉出来需要能量一样，光子也需要消耗能量以使自己远离大质量天体附近的区域；从在地球引力范围内移动的光子当中，人们也已经测量到了这种效应。光子的能量与其波长成反比：高能光子波长短，而低能光子波长长。[38] 光子在从黑洞逃离时会失去能量，因此其波长会增加。这会改变光的颜色，使其

在光谱上从蓝色端（短波）向红色端（长波）移动。这种移动被称为**引力红移**，产生于时空本身的延展，或由于黑洞之类的大质量物体作用而弯曲的地方。要注意的是，约翰·米歇尔虽然在暗星问题上给出了重要的猜想，却错误地认为当光从势阱中爬出时速度会降低。我们现在知道，受大质量恒星影响的是光的波长，也就是频率。

黑洞附近的时间会受到什么影响

在第1章和第2章中，我描述了时空是如何因质量（也就是自身会产生引力场的物体）的存在而变形的，这意味着在黑洞附近，不仅是空间，时间也会受到影响。

想象一下，你想与史瓦西黑洞保持安全的距离，但是你又想了解在其附近时间是如何表现的。因此，你安排了26名固定的观察者安全驻扎在黑洞外靠近事件视界的地方。这些观察者按照从A到Z的顺序命名，并且排成一条直线，其中A最接近事件视界，而Z最接近安全地待在远处的你。从A到Z的每个观察者都有一个精确的时钟，来测量他们所在的特定位置的时间。为说服A到Z参与这个实验，你还为他们每个人额外提供了一个不同寻常的时钟作为礼物。这些时钟经过调整，与你所在的安全位置的时钟读数相同。最接近你的参与者Z会发现他所拥有的两个时钟所读取的时间略有不同，因为他自己的时钟测量的是当地时间（术语叫作“**固有时**”），会比在更远更安全的距离所测量到的时间更慢一点。参与者Z到A所整理出的结果将显示出一个显著效应：与特别调整过的礼物时钟上所显示的远处时间相比，越接近黑洞的测时时钟“运行得越慢”。爱因斯坦的广义相对论所描述的这种效应被称为**时间膨胀**。对于更加靠近黑洞的观察者来说，效应会越来越明显。与远处观测者使用的时钟相比，本地时钟（不管是原子钟还是生化钟）越靠近黑洞，运行速度就会越慢。

39

假设让另一组26名观测者同样站在一个不同的黑洞旁，进行实验的另一项任务。他们的排列方式与第一个黑洞附近的同名观测者相同。不过在第二种情况下，黑洞的质量是第一次实验中黑洞的两倍。相比第一个实验，你为第二组观察者所准备的礼物时钟进行了彻底的调整，直到每个礼物时钟的速率恰好是第一组实验中对应时钟的两倍。因为它们到黑洞中心的距离与第一组完全相同，而第一个黑洞的质量仅为第二个黑洞的一半。黑洞质量越大，时间膨胀效应也会越大，而且越接近事件视界，这种效应会变得越极端。

请注意，这种时间膨胀不是因为时钟离黑洞更近，而离你这个远处的安全观测者更远造成的。光有额外的传播时间，对于远离黑洞的观察者来说，仅仅补偿传播时间是不够的。无论你用哪种可靠的方法，时钟越接近黑洞，所测得的时间流逝的速率就越慢。时间本身被拉长了（或者说膨胀了）。

40 黑洞附近时间膨胀的必然结果是什么？在黑洞附近的观测者的参考系与离黑洞非常遥远的观测者的参考系中，其效应产生的结果截然不同，实际上可以说是天差地别。

现在让我们思考一下，在你的第一个实验中，如果观察者 A 变得有点粗心，弄掉了他的第一个时钟（就是测量固有时的那个），并使它落向黑洞，将会发生什么。尽管发生了这场灾难，他仍然可以紧紧抓住你诱使他参与实验的礼物时钟。你和 A 都会看到他的第一个时钟向洞口移动。时钟会发现自己越来越快地进入黑洞。你和 A 会注意到，在下落的时钟上读到的时间与 A 的礼物时钟（被调整为比本地时钟运行得更快，与你的时间相同的时钟）上的时间差异更大了。过了一会儿，你和 A 都会注意到下落时钟的读数停止了。从事件视界向远处的观测者发出的光子似乎无限期地停在了那里。任何落入黑洞的物体在穿过事件视界的临界半径之后所发生的事情，对外部的观测者来说都是不可知的。因此，事件视界可能会被当成时空中的一个洞。正如我们在第 1 章中看到的那样，光无法从事件视界中逃出，这就是为什么事件视界是黑色的。然而，在通过事件视界直线下落的时钟的参考系中，生活远非一成不变。从时钟的角度来看，假设黑洞的质量是太阳的 10 倍，那么它将在仅仅十万分之一秒内落到奇点。如果时钟不幸落入一个质量是太阳 10 亿倍的超大质量黑洞（就像我们在第 8 章讨论类星体时将遇到的那个），那么它在极大的事件视界和奇点之间的旅程将会是悠闲的几个小时。

黑洞附近的潮汐力

41

假设 A 一时心软，希望与他掉落的时钟重新团聚，并想知道脚朝下跳进黑洞会发生什么。接下来的事可以证明这种跳跃将是一个严重的错误，因为他最后生存的概率将为零。作用在他脚上和头上的引力之间的差异将会变得非常大，大质量物体的引力场都有这种特征，它们都是平方反比例场。地球与月球的距离很远，但即使这样，地球两侧受到的月球引力的微小差异（也就是所谓的潮汐力），也会导致每天两次的涨潮落潮。一般来说，由不同位置的引力差异所导致的力都被称为潮汐力。还有其他因素丰富了潮汐涨落的细节，例如月球的相对角度产生的引力以及大陆板块的具体形状。但即使地球表面完全被海洋覆盖而没有陆地，仍然会有潮汐，其导致的海平面每天两次的变化幅度约 20 厘米，这仅仅是因为地球上与太阳距离不同的点所受到的引力不同。

现在让我们来考虑一下比我和地球中心之间更小的距离。当我坐着写下这一章的时候，我的头比我在书房地板上的脚要高出 1 米多。因此，我的脚比我的头更靠近地球的中心。因为引力遵循平方反比定律，所以会表现得好像地球的所有质量都集中在地球的正中心；而因为我的脚比我的头到这个中心的距离更小，所以我的脚会感受到更大的引力。但实际上，这个差异是相当微弱的：每相隔 1 米，重力的差异是千万分之三。这种差异这么小是因为我距离地球中心大约 6400 千米。接近黑洞这样的点质量时，径向相隔 1 米的两个点所受到的引力差异将更加极端。在接近

42

奇点时，A 的脚所受到的拉力将会超过他的肌腱和肌肉的承受范围，从而将它们从 A 身上撕扯下来，而他本人会被拉长成类似长意大利面一样的东西。所以，最好不要跳向黑洞。

动态时空

黑洞的旋转不仅对围绕它运行的物质的轨道有着重要影响，还可以影响从黑洞中提取能量的多少。根据罗伊·克尔的工作和他的爱因斯坦场方程解，我们知道了粒子绕着黑洞转的最小轨道只取决于黑洞的旋转速度。如图 13 所示，黑洞旋转的速度越快，物质在被黑洞吞噬之前就能靠得越近。尽管黑洞外面除了虚空什么都没有，但如果你让某个东西直直落向一个旋转的黑洞，它将开始绕着这个黑洞转。在能层之外，还有可能靠火箭克服这种参考系拖曳，但在它的内部则不能这么做。在旋转黑洞能层内部，也就是事件视界之外的区域，没有任何东西可以停滞不前。旋转黑洞实际上拖曳了时空，以及周围时空中的物体。这种参考系拖曳的另一个效应是，即使光正朝着黑洞旋转的反方向行进，它也将被带动以相同的方向绕着黑洞转。

43

图 13　旋转黑洞轨道上的气体会比无旋转黑洞轨道上的更靠近黑洞

绕着黑洞运行

思考一下，如果我们的太阳在此刻突然变成一个黑洞，会发生哪些事情，这非常有意思。你或我注意到的第一件事将会发生在 8 分钟后：我写作时所沐浴的美丽的春日阳光会突然消失。虽然与第 8 章将会讨论的类星体和微类星体相比，我们称之为太阳的这颗孤单的恒星的光度很小，但它离地球足够近，可以为我们的星球提供平均每平方米 1 千瓦的能量。值得注意的是，这足以供养地球上的所有生命：植物得以生长，然后被动物吃掉，再然后动物又被其他动物吃掉。太阳一直是这一切背后的推手。如果太阳内部的聚变停止，它会（出乎意料地）坍缩成黑洞，那么它将变得非常黑暗，我们最终都会死去（这是一个有点悲观的前景，但我希望读者坚持看到第 7 章，在那里我们将了解到太阳不是那种能形成黑洞的恒星——它太轻了）。但从动力学角度而言，对于地球这颗行星和我们所考虑的整个太阳系的行星、矮行星和小行星来说，什么都不会改变。绕着太阳运行的所有大质量物体将在几乎相同的轨道上继续运行。引力的作用方式是，无论太阳是否具有与现在相同的大小，或者是否坍缩成直径 3 千米的事件视界以内的奇点，其外面的引力都将保持不变。在引力作用下，球对称坍缩成的黑洞并不会改变绕转物体的角动量，因此太阳系内的规律、演化和潮汐都将完全不变，只是缺少了阳光。 44

但是在之前充斥着太阳等离子体的地方，会存在一些更接近太阳形成的黑洞的新轨道。不过这些轨道不能离事件视界太近。大质量奇点使时空弯曲，一些细节的变化意味着在紧挨着事件视界的地方绕转是不可能的。试图沿着圆形轨道绕转就需要用火箭进行修正以维持轨道。实际上，数学计算表明，对于我们或任何有质量的粒子来说，可能存在于静态黑洞附近的稳定圆形轨道的最小半径是史瓦西半径的 3 倍。正如前文所说。

实际上，不稳定的圆形轨道距史瓦西（无旋转）黑洞的距离可能会达到这个距离的一半。这个距离定义了一个有时被称为**光子球**的球面。即使对于光子来说，这些轨道也是不稳定的，并且在不久之后，绕转的光子将要么向黑洞偏离再也回不来，要么就跑到太空中去了。

对于存在自旋的克尔黑洞而言，黑洞附近的轨道会有所不同。特别是与静态的史瓦西黑洞周围只有一个光子球不同，这里会存在两个光子球。最外层的球面上是与黑洞旋转方向相反（也就是我们所说的在**逆行轨道**上）的光子。在其内部的光子球上则是与黑洞旋转方向相同（在**顺行轨道**上）的光子。对于那种与史瓦西黑洞并没有太大不同的自转非常缓慢的黑洞，这两个光子球在空间距离上非常接近。随着黑洞的自旋越来越快，两个球面也将越来越远。

45

在离旋转黑洞更近的地方，还有着另一个重要的界面（在第 3 章中讨论过），被称为稳态极限面。在一个遥远的观测者看来，没有任何东西可以在这个表面上保持静止——就算你装备的火箭无比强大，都不可能在离旋转黑洞这么近的地方保持不动。在这个界面上，即使是反向（转动）的光线也会被拖成顺着旋转方向转动。虽然仍然可以借助足够的推力从如此靠近旋转黑洞的地方逃离，但是在这里任何东西都不可能保持静止不转。继续向内前进，下一个重要的表面就是我们在第 1 章中讨论的事件视界，这也是我们最初在史瓦西黑洞情况下遇到的单向膜。和静态黑洞的情况一样，向外穿越事件视界是不可能的，而向内穿越它所面临的命运则不可避免地和向内穿越无旋转黑洞一样。

克尔黑洞周围的轨道一般不会被限制在某个平面上。只有那些落在赤道面上的轨道才是被限制在平面上的轨道（与旋转黑洞镜像对称的平面）。这个赤道面外的轨道会在三维空间中移动。这些轨道被限制在由最大和最小半径以及被赤道平面的最大角度所限制的范围内。

黑洞自旋这一细节对粒子能离黑洞多近有着显著影响，并且取决于粒子相对于自旋的行进方向。对于极限自旋的黑洞，轨道与黑洞自旋同向（正旋）的光线对应的光子球的半径是史瓦西半径的一半。对于逆行轨道上的光线，其光子球的半径是史瓦西半径的两倍。对于顺行轨道上的有质量粒子，它们可以绕转的最内稳定圆轨道也是史瓦西半径的一半。对于逆行轨道上的那些粒子，这么近的距离将是不稳定的，它们的最小稳定圆形轨道是史瓦西半径的 4.5 倍。因此，

46

相比无旋转的黑洞，旋转的黑洞可以让顺行轨道上的粒子在更靠近黑洞的轨道上运转，只要还没到事件视界，粒子就有回头的余地，否则将无法返回。在第 7 章中，我们会考虑这两件事的重要性：物质在落入黑洞之前能绕着多靠近黑洞的轨道运转，以及可以从中提取多少能量。

05

黑洞的熵和热力学

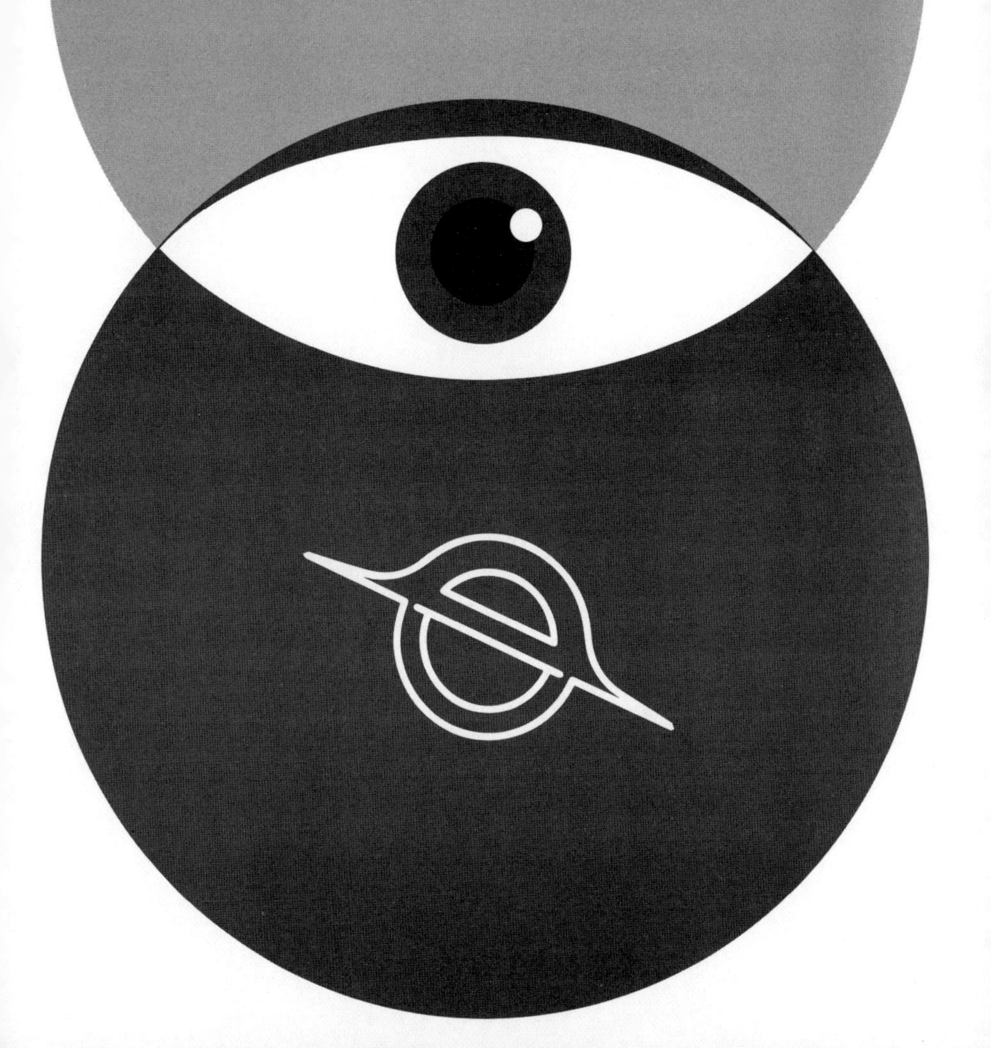

人如其食

47

人们常说人如其食。如果你只吃垃圾食品和巧克力，那么你的气色和身心健康都将与你食用沙拉和地中海式饮食这类健康食品时大不相同。但是，黑洞似乎并不挑食。无论是吸收广阔的星际尘埃还是吸收一整个立方光年的煎蛋，它们的质量都会无可避免地增加。实际上，在黑洞吃完丰盛的食物后，你无法分辨它吃了什么，只能知道它吃了多少（不过你可以知道它吃的东西是否带有电荷或角动量）。你只知道饮食的数量，而不是饮食的品质。第2章中介绍的“无毛定理”说，黑洞仅有很少的参数（质量、电荷和角动量）表征，因此我们无法讨论黑洞是由什么构成的。

对黑洞所吸入物质的性质缺乏了解，这似乎是个微不足道的事情，但实际却有深远影响。有关黑洞午餐菜单的信息从最开始就丢失了。落入黑洞的任何物质都已放弃了自己的特性，我们无法对它进行探测，也无法了解关于它的任何细节。

黑洞与引擎

48 对于那些研究过热力学这门美丽的学科的人来说，以下情况再熟悉不过了。在该领域，理解信息是如何通过物理过程丢失或耗散掉是很容易的。热力学有着悠久而有趣的历史。关于热力学的现代理论始于工业革命，当时人们试图研究如何提高蒸汽机的效率。在对“能量”进行定义时，应要求其始终保持守恒，并且可以在不同形式之间进行转换，这被称为热力学第一定律。尽管你可以让能量在不同类型之间进行一些转换，有些特殊的转换却是不被允许的。例如，尽管你可以将机械能完全转换为热能（每当你踩刹车使汽车完全停住时都在这样做），但你无法将热能完全转化为机械能；不幸的是，这正是我们想用蒸汽机做到的事情。因此，火车上的蒸汽机只能将炉子中的热量部分地转化为使车轮转动的机械能。人们最终意识到，热是一种涉及原子随机运动的能量，而机械能则涉及一些诸如轮子或者活塞这种大块物质的协同运动。因此，热的本质中的一个重要部分就是随机性：由于物体内原子的振动，你无法跟踪单个原子的运动轨迹。这种随机运动不可能在没有任何额外代价的情况下被非随机化。在任何孤立系统的各种物理过程中，这种随机性（学名叫作**熵**）都不会减少，且必须始终保持不变或增加——这就是热力学第二定律。对这种现象的一个解释是，由于无法跟踪大型系统中所有原子的运动，我们所知的关于世界的信息总是在减少。随着能量从宏观尺度转移到微观尺度，也就是从简单的活塞运动转化为大量原子的随机运动，对于我们来说信息就丢失了。热力学使我们能够将这个模糊的概念完全定量化。事实证明，这种信息的丢失与我们所描述的物质落入黑洞是完全类似的。

49

尽管热力学是为蒸汽机发展起来的，但这些原理被认为适用于宇宙中的所有过程。最早认为这与黑洞有关的人之一是牛津物理学家罗杰·彭罗斯。他认为由于黑洞有自旋，我们有可能从中提取能量并将其用作某种引擎。他提出了一个巧妙的方案，将物质投向一个旋转的黑洞，使该物质的一部分带着比被扔进去时更多的能量跑出来。能量是从事件视界之外的区域提取的（实际上就是从第 3 章中讨论过的能层提取的）。彭罗斯过程减慢了黑洞的转速。原则上，可以通过这种方式从黑洞中提取大量的能量，但这当然还只是个思想实验，因此目前似乎还不能用它来解决地球这颗行星上迫在眉睫的能源危机！在彭罗斯的工作完成几年之后，詹姆斯·巴丁（James Bardeen）、布兰登·卡特（Brandon Carter）和斯蒂芬·霍金取得了划时代的进步，并用公式表示出了黑洞动力学三定律，这为霍金后来对黑洞热力学的思考奠定了基础。它用到了由黑洞的质量和自旋决定黑洞的温度这一概念。

黑洞和熵

彭罗斯的洞察力是促使其他人思考黑洞热力学的一个重要因素。在弗洛伊德（R. M. Floyd）的帮助下，彭罗斯展开了想象，在他脑海里，黑洞事件视界的面积趋于增加。斯蒂芬·霍金开始研究彭罗斯巧妙的方案。事件界的面积以一种相当复杂的方式依赖于质量和自旋（和电荷），但是霍金证明，在任何物理过程中，这个面积始终会增加或保持不变。一个有趣的效应是，如果两个黑洞合并，则合并后黑洞的事件视界面积大于两个原始黑洞的事件视界面积之和（这在直觉上是可靠的，因为事件视界的半径与质量成正比，而众所周知表面积与半径有关）。这与我们在热力学中所看到的熵的情况相同，因此人们开始怀疑是否黑洞的熵和它的面积有着某种联系。这不仅仅

50

是一个有趣的类比，不是吗？约翰·惠勒的一个学生雅各布·贝肯斯坦（Jacob Bekenstein）走在了前面，他在自己的博士学位论文中提出了一个直接的联系。贝肯斯坦运用热力学中信息论的观点，论证了黑洞事件视界的面积与它的熵成正比（这一选择意味着你要将事件视界的面积除以普朗克面积，并在乘以一个数值因子后得到熵。普朗克面积是一个物理学基本常数，它的量级是 10^{-70} 平方米。选择这个单位会使黑洞的熵特别大）。

最初，霍金并不相信贝肯斯坦的研究结果。但在进一步的验算中，他不仅证实了这个结果，而且加深了我们对黑洞热力学原理的理解。或许我们应该去了解如何进行这些分析，这样我们既可以理解它的优势，也可以理解它的局限性。研究该领域的理想方法，是使用结合了量子力学和广义相对论的、被称为量子引力的方法，用它来研究类似黑洞中的奇点这种非常小但引力在其中有着重要作用的系统，比如黑洞中的奇点。不幸的是，我们目前还没有一个很好的量子引力理论。一个还不错的方法是使用广义相对论来计算时空如何弯曲，然后将其与量子力学一起使用，以理解粒子在弯曲时空中的行为。这就是霍金试图理解黑洞热力学的方法。

真空是空的吗

51　真空（也就是什么都“没有”的区域）这一概念有着悠久而曲折的历史。大多数古希腊哲学家都痛恨这个概念，理由在今天看来非常神秘，但还是有一小群原子论者将真空纳入了对世界的描述。因此在科学复兴之前，真空的想法已经非常过时了。但随着1650年人们发明了空气泵，真空可以通过实验被证实。尽管按照现代的标准，在17世纪，从容器抽出空气所能够提供的真空度仍然很差，但虚空的观念已变得更可信了。随着人们在20世纪初期证明了原子的存在毋庸置疑，验证某个空间区域中有没有原子，不仅变得无可争议，而且不可避免。

原子的存在被证明不久后，就出现了新的物理学理论——量子力学。这种新理论的一个令人惊讶的结论是：在短暂的瞬间内能量似乎不需要守恒。热力学第一定律是物理学中最重要，而且看上去牢不可破的原理，它坚称无论何时何地，在能量的借方和贷方之间都必须进行严格的核算。“能量必须始终保持平衡！”宇宙的会计大声疾呼。实际上，宇宙的会计规则似乎更加宽松，并且有可能获得信贷。在短时间内借用能量是完全可以接受的，只要你随后迅速偿还即可。你所能借到的金额取决于贷款的期限，而这个量则由海森堡不确定性原理所描述。例如，即使在所谓空无一物的真空中，也可以借用足够的能量来产生粒子和反粒子对。这两个物体可能在一瞬间产生，并在持续极短时间后湮灭，而后在所允许的最长时间限内偿还能量（时间间隔越短，借入的能量就越多）。这样的过程每时每刻都在进行。我们甚至可以测出这个过程！现在，我们知道了真空实际上不是空的，而是由这些成对产生并消失的虚粒子构成的场。因此，真空不是空无一物的不毛之地，而是充斥着量子层面的活动。

52

黑洞蒸发和霍金辐射

霍金使用现代关于真空的理论，也就是量子场论来研究粒子在黑洞事件视界附近的行为。他的分析是数学化的，但我们可以用一种非常简单的方式来描述它。实际上，在黑洞的事件视界附近产生的一对虚粒子，也就是一个粒子及其反粒子（电荷相反，质量相同），最后可能被拆散。如果这对正反粒子中的一个落入事件视界，它将陷入奇点，并且永远无法恢复。

但是，其伙伴可能仍留在黑洞之外。这个粒子失去了它的虚拟伙伴，但现在是一个真实的粒子，并有逃逸的可能。如果粒子确实逃了出来而不是掉了回去，那么它就成了霍金辐射的一部分。在远处的观察者看来，黑洞已经因为发射粒子而损失了质量。人们已经认识到，在考虑量子场论的情况下，黑洞并不是完全黑的，它可以辐射出粒子。这个论证也适用于光子。所以，如果霍金的论点是正确的，黑洞会发出非常弱的光（也被称为电磁辐射）。

所有非零温的物体都会以光子的形式发出热辐射。你本人也会这样，这就是为什么即使在黑暗中你也会出现在红外摄像机上（这也正是警察和军方使用该类摄像机的原因）。物体越热，辐射的频率就越高。人体只能发出红外辐射，但红热的火钳可以热到发出可见光。因为黑洞会发出霍金辐射，所以如前所述，它会具有一个温度（被称为霍金温度），然而通常这一温度非常低。质量为太阳 100 倍的黑洞，它的霍金温度比绝对零度（比水的冰点低 273 摄氏度）只高不到十亿分之一摄氏度！这就是霍金辐射尚未被检测到的原因之一：它太弱了。但是，人们相信它确实是存在的。

53

然而，霍金辐射确实对黑洞的演化产生了一个有趣的影响：它是导致黑洞最终死亡的罪魁祸首。再想想这两个虚粒子。从黑洞中逃出的实粒子的能量必须为正，但由于虚粒子对是从真空中自发出现的，吸入黑洞中的虚粒子必须具有负能量作为补偿。因为能量和质量是相关联的，所以这个过程的效果是黑洞净增加了负质量。因此，由于发出霍金辐射，黑洞的质量将会降低。

由此，霍金发现了一种可以令黑洞蒸发的机制。随着时间的流逝，黑洞会慢慢发出辐射并损失质量。最初，这个过程非常缓慢。事实证明：黑洞越大，其“表面引力”越小。这是因为尽管表面引力取决于质量，但仍遵循平方反比定律。对于更大的黑洞来说，质量会更大，体积也更大。因此最终结果是，大黑洞的表面引力会很小，而这相当于温度非常低。因此，大的黑洞比小的黑洞发出的霍金辐射更少。

但是，随着黑洞蒸发并损失质量，霍金辐射的量会随着表面引力的增加而上升，因此温度也会升高。 54 假设黑洞没有吸收任何其他的能量，这种效应将使质量损失的速度越来越快，直到黑洞在寿命尽头突然消失。因此，黑洞的寿命结束时并不会发出一声巨响，而是更加安静，发出“砰”的一声。这种蒸发过程仅适用于温度高于其周围环境温度的黑洞。在宇宙历史的当前阶段，从宇宙微波背景辐射的光谱形状测得的宇宙温度比绝对零度高 2.7 摄氏度，因此质量超过 10 万亿千克的黑洞现在不会蒸发，因为它们的温度低于其周围环境的温度。但是，当宇宙随着进一步膨胀而变得更冷时，这些质量比太阳小得多的黑洞就能蒸发了。在宇宙时间的这个时候，宇宙中所有比太阳质量的百分之一还要小的黑洞都应该已经消失了。

黑洞信息悖论

这一切会引发的一个问题是：落入黑洞的物质中储存的信息会怎么样？一种观点认为，即使黑洞在物质落入后马上就蒸发掉，物质中的信息也会永远丢失。另一种观点则认为那些信息不会丢失。后者认为，因为黑洞蒸发了，所以落入黑洞的原始物质中所包含的信息必须以某种方式存储在黑洞的辐射中。因此，如果你可以分析来自黑洞的所有霍金辐射并完全理解这些辐射的意义，你就能重建最初掉入黑洞中的所有物质的细节。关于这件事，在斯蒂芬·霍金、基普·索恩（Kip Thorne）和约翰·普雷斯基尔（John Preskill）之间有一个著名的赌注。索恩和霍金支持前一种观点，而普雷斯基尔支持后一种观点。他们的赌注是，败者将输给胜者一套他所选择的百科全书。2004年，霍金被“信息可以被编码在黑洞的辐射中”的想法彻底说服了，于是他认输并给了普雷斯基尔一套关于棒球的百科全书。不过，这件事现在仍然有争议。

55

尽管这些理论推测都很巧妙，但值得再次说明的是，人们连黑洞发出的最普通的霍金辐射都还没有观测到。物理学的历史上充斥着古老精巧但最终被证明是谬误的理论残骸。实验和观测经常会产生出乎意料的结果。确实如此，从我们对宏伟天文现象的观测可知，或许根本没人预测出黑洞的基本原理。未能观测到这些微弱的霍金辐射的原因之一，是我们所知道的许多黑洞都位于宇宙中某些最亮的物体的中心。而这些黑洞实在太大了，也就是说它们都太冷了，不能通过霍金辐射蒸发。这些物体由于完全不同的原因而异常明亮，我们将在第6章和第8章中对此进行探讨。

06

怎样给黑洞称重

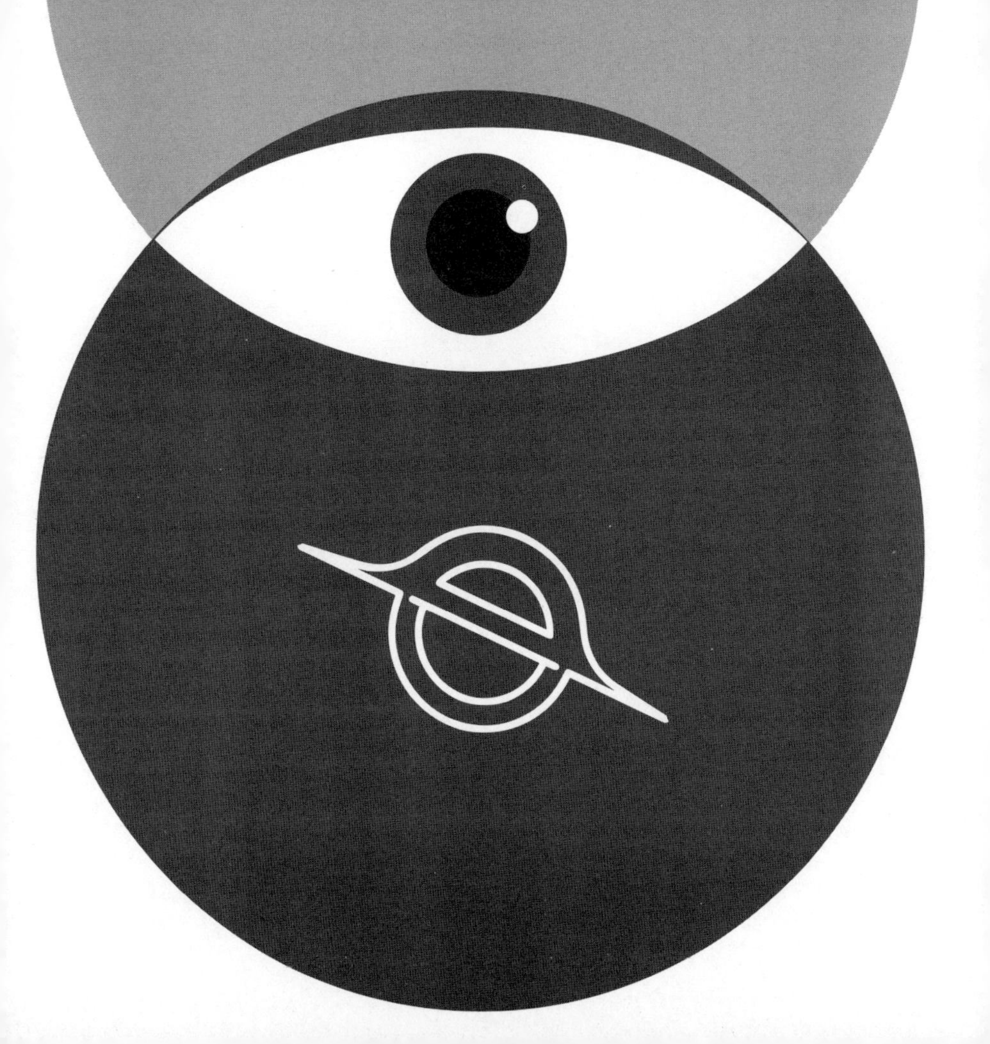

56

太阳以及围绕它运行的行星、矮行星（冥王星是其中最著名的例子）、小行星和彗星共同组成了太阳系。太阳系本身在银盘上绕着位于银心的质心转动。我们的太阳系在银盘中的圆轨道上绕转的速度大约是7千米/秒，绕着银心转完一整圈需要几亿年。除了这种轨道运动，整个太阳系还垂直于**银道面**运动。它所表现出的这种运动是物理学家所熟知的简谐运动，而把我们的太阳系拉回到位于银道面上的平衡位置的回复力，则来自构成银盘的恒星和气体的引力。目前我们在这个平衡点上方约45光年的地方。从现在起大约2100万年后，太阳系将到达银盘上方320光年处的极值点；在此之后再过4300万年，太阳系将重新回到银河系的中心平面。当太阳系位于银道面的中心时，地球将最大限度地暴露在宇宙射线中。这些宇宙射线会在银道面上呼啸着转圈，它们被磁力线所俘获，以介于完全杂乱无章和完全有序之间的某种方式运动——一边沿着磁力线前进，一边绕着磁力线旋转。有人猜测可能是由于太阳穿过银道面的运动导致了恐龙灭绝。但是这种推测很难被证实或反驳，因为

57

这种轨道运动的时标对于寿命通常不会超过一个世纪的人类观测者来说显然是难以观察的。人们采用足够精确和彻底的手段进行天文学观测也只有几个世纪。因此，在天文学观测中，当我们想要关注某种以更长时标变化的过程时，这是一个很常见的问题。

然而，至少在所用时标与人类及其望远镜所关注的时间尺度差不多的情况下，银河系中的轨道运动非常容易测量。既然我们在讨论黑洞，那么最令人感兴趣的显然是银河系最内部区域中恒星的轨道运动，这一区域位于天空中被称为人马座 A* 的那一部分。当我们观察这个在南半球最容易看到的区域时，也是在看向距我们 27 000 光年远的银河系的正中心。这是一个天体特别稠密的空间区域，而当我们想研究银心时，会面临两个问题：首先是恒星的空间密度较高，其次是尘埃很多。

第一个问题意味着你需要使用一种能够实现高分辨率成像的测量技术，也就是精细的细节可以被区分开，就像在给定的相机上长焦镜头提供的细节比广角镜头提供的细节更精细一样。仅仅使用更大的望远镜肯定不足以解决这个问题，因为除非我们把望远镜放在大气层外的卫星上，否则我们将不可避免地通过具有湍流的大气来观察所有的天体。不过，人们已经开发出了各种各样的技术来消除地球大气中湍流的影响。最为重要的是一种被称为**自适应光学**的技术。这种技术可以校正大气变化的观测误差，其原理是对准明亮恒星（又叫导星）并调整望远镜的主镜，直到模糊的图像变清晰。当所感兴趣的天空中没有明亮的恒星时，则可以向上发出高功率的准直激光束，以激发大气中的原子，并由此进行大气校正。

58

第二个问题是朝着银心的方向存在着大量的星际尘埃，这导致的问题是：很难透过尘埃看到可见光，就像来自太阳的紫外线很难透过不透明的遮阳帽一样。解决这个问题的方法是：我们需要在红外波段而不是在可见光波段进行观测。

如何测量银河系中心的黑洞的质量

这种红外观测得到了两个小组的支持，一个小组由美国加利福尼亚州的安德烈·盖齐（Andrea Ghez）领导，另一个小组由德国的莱因哈特·甘泽尔（Reinhard Genzel）领导。两支团队的工作均独立提供了对银河系中心质量的非常精确的测量结果。图 14 展示了安德烈·盖齐和她的团队的数据。在过去的几年中，他们对银心的中心区域进行了多次观测，并观察恒星自上次观测以来的移动情况。因为这些恒星的光谱类型是已知的，因此它们的质量也是已知的。年复一年，随着每颗恒星的轨道路径变得清晰，盖齐及其团队能够根据动力学方程（开普勒定律，也是主导了太阳周围行星运动的定律）独立求解每个轨道，并推算出这些轨道共同的焦点所在的“黑暗”区域的质

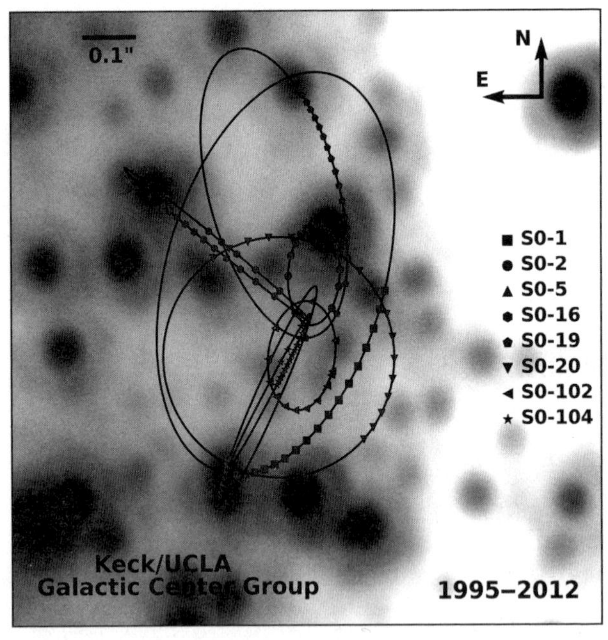

图 14　绕银河系中心黑洞运动的恒星的连续位置[1]

1　图上文字为：凯克望远镜/加州大学伯克利分校银河中心小组。

量。这些独立的解很好地确定了该暗区的质量。现在人们知道，暗区在半径不超过 6 个光时的区域内，具有的质量刚好是太阳质量的 400 万倍。因为这个物体虽然不可见，但质量非常大，所以唯一的结论就是我们银河系的中心存在一个巨大的黑洞。

59 没有理由认为我们所在的星系——银河系是唯一一个中心存在黑洞的星系。相反，人们强烈怀疑所有星系，至少在更大的星系的中心，都可能存在一个黑洞。这是由于当时在杜伦大学的约翰·马格里安（John Magorrian）和同事发现了一对看起来非常基本的关系，也就是星系中心的黑洞质量与星系质量的关系。当然，不论测量黑洞的质量还是星系的质量都非常困难。在我们银河系中心表现得如此出色的技术无法应用于系外的星系，因为它们太远了。

60 椭圆星系中心的黑洞质量超过了太阳质量的 100 万倍，实际上可能达到甚至超过太阳质量的 10 亿倍。因此，它们通常被称为**超大质量黑洞**。

尽管在测量黑洞质量和星系质量方面存在着困难，但是人们已经发现，在各种不同的星系中，中心黑洞的质量与其宿主星系的质量成比例。人们认为这暗示着中心黑洞和星系本身在整个宇宙的时间尺度上是协同生长和演化的。

银盘上遍布着许多黑洞

除了位于银河系中心的单个中央超大质量黑洞外，人们认为在每个星系的范围内还散布着几百万个黑洞，并相信这些黑洞与星系中心黑洞的形成方式非常不同。星系中心的黑洞是通过逐渐吸收下落的物质增长，而这些**恒星质量黑洞**的前身则是大质量恒星，曾经发出过非常明亮的光芒，其内部的聚变产生了能量并使之保持高温高压，最重要的一点在于这些能量可以抵抗引力坍缩。当它们的核燃料全部耗尽时，就不再存在可以支撑恒星的辐射压，因而就没有任何东西可以平衡向内的引力。对于质量与我们的太阳相似的恒星，在引力作用下的坍缩最终会形成一个被称为白矮星的致密物体。“致密”一词在天体物理学中具有特殊的含义，表示该物质的密度与普通物质完全不同。按照普通物质密度的标准，白矮星中的物质已经被极度压缩，所以它是致密的。这些物质的所有电子都与其原子核相分离，也就是被电离了，但是又很冷（通常物质仅在高温下才会被电离）。电子产生抵抗持续向内引力的压力，是因为它拒绝被压缩到过于狭窄的区域（这是海森堡不确定性原理的结果），这种效应的学名叫电子简并压力。恒星会在用尽所有燃料

61

后坍缩，如果恒星的质量更大一些，那么使物质收缩的引力也会更大，使得电子和对应的质子融合形成中子。这样就可以形成比白矮星更加致密的天体——中子星。

如果我们对黑洞感兴趣，那么我们必须转向比那些会生成白矮星甚至中子星的恒星质量更大的恒星。质量更大的恒星在其燃料持续存在且核聚变能够维持的期间将会非常亮。一旦所有燃料都被用完，恒星的寿命就结束了，发出的光也会熄灭。如果这颗恒星现在已经足够重，以至于引力甚至可以压垮强大的“中子简并压力”，那么因此导致的坍缩会强到连中子简并压力也无法平衡，于是坍缩就会不可避免地导致黑洞的产生。大质量恒星的坍缩通常伴随着壮观的超新星残骸的爆发，而在原来恒星的位置上只会留下一个黑洞。在这样的爆炸中，许多元素，尤其是比铁重的元素，都被合成了出来。

第一个通过测定双星系统中两颗星的质量而认证的黑洞叫 V404 Cyg。豪尔赫·卡萨雷斯（Jorge Casares）与菲尔·查尔斯（Phil Charles）和他们的同事非常仔细地观测了两颗星的轨道，并从分析中推断出这对双星包含一颗质量至少是太阳 6 倍的致密星，因此它就是一个黑洞（后来发现它的质量其实是太阳的12倍）。

我们能对银河系中的恒星数量及其质量进行合理估算，然后估计出有多少大质量恒星出现的时间够早（这样它们就通过聚变用完了所有的核燃料），这样就可以估算银河系中恒星质量黑洞的数量。即使我们银河系中只有极少比例的恒星会演化成黑洞，但因为银河系中有超过 10^{11} 个星体，所以我们仍然会有许多黑洞。

62

我们如何测量这些遍布星系的黑洞质量？实际上，对于某些恒星所残留的黑洞，需要用到的技术在动力学上与测量银河系中心的黑洞时所使用的技术非常相似。原因是银河系以及其他星系中的很大一部分恒星，都形成了成对的双星系统。我们很容易猜测到这是怎么发生的：引力使得物体互相吸引，而很多两体轨道都是稳定的，因此一旦两颗恒星相遇并被引力束缚在一起，它们就很可能会保持这种状态。对于双星系统，如果我们可以测量恒星彼此绕转完整一圈所花费的时间——也就是轨道周期的时长，并且如果我们知道它们之间的距离，那么就可以知道它们的质量。如果致密星绕着光谱类型与质量都已知的正常恒星（正在发生聚变）的轨道运动，那么致密星的质量就很容易测出。如果类似黑洞这样的致密星是孤立的，没有处于双星系统中，那么缺少其动力学信息就意味着没有办法推断出它的质量或确定它确实是黑洞。我们可以测量的最小的黑洞质量是太阳的几倍，但是最重的恒星质量黑洞可能比我们的太阳重100倍。

在当前的技术条件下，测量黑洞质量非常容易，不过这仍然需要良好的耐心和韧性。鉴于质量本质上只是黑洞两个基本的物理特性之一，因此这些研究只能让我们了解它一半的特性。不过，测量黑洞自旋的难度要更大，在第7章中，我将会描述尝试并做到这一点所需的英勇努力。

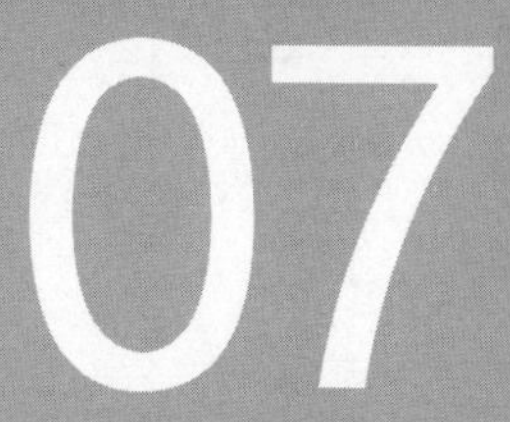

吃得更多，长得更大

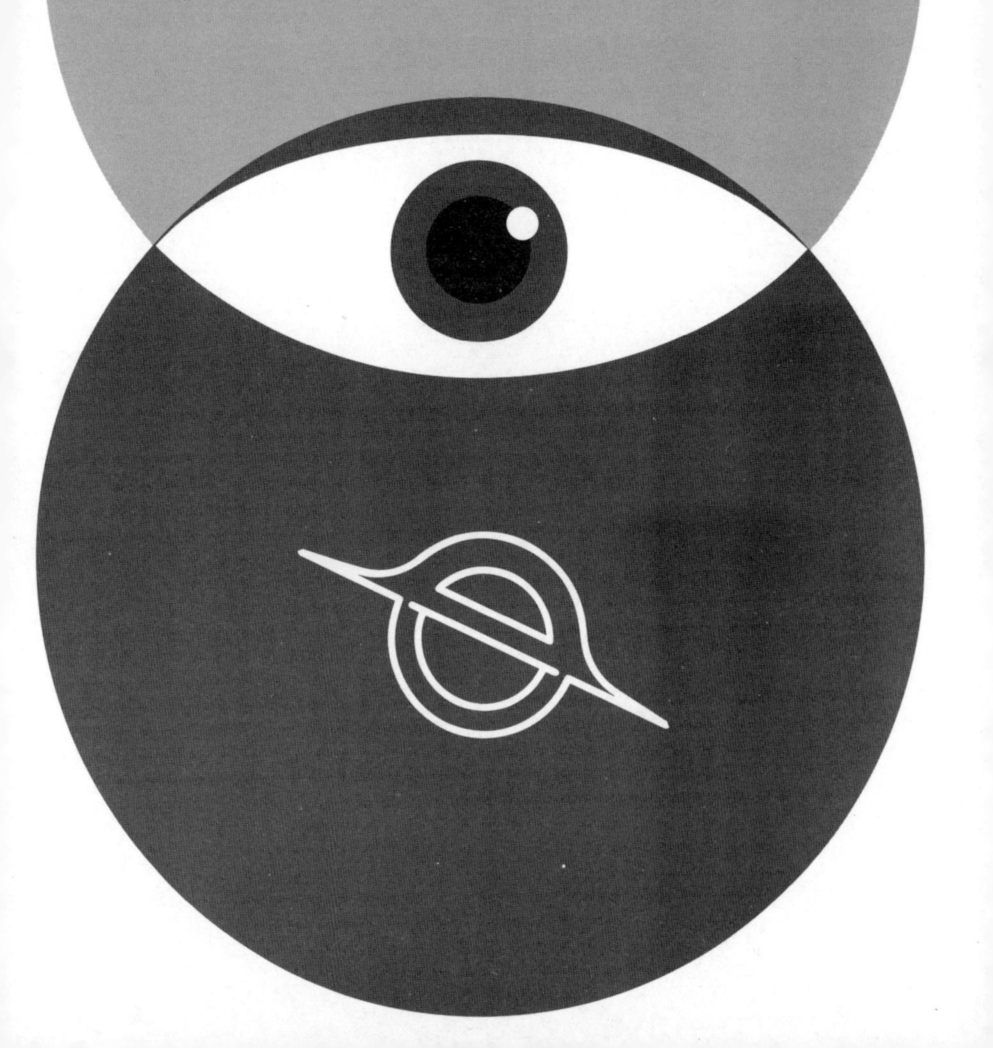

它们吃得有多快

[63] 黑洞会“吞噬其周围一切物质”这一流行观点，只在事件视界附近成立，并且坠入物质的角动量还不能太大。在远离黑洞的地方，其外在引力场与质量相同的球形物体的引力场相同。因此，粒子可以按照牛顿动力学原理绕黑洞公转，就像它绕着其他恒星公转一样。是什么打破了粒子划着圆圈（实际上是椭圆）不停转下去的模式，让它按照更奇特的轨迹运行呢？答案是，总是有不止一个粒子在绕着黑洞转。我们观察到的天体物理学现象之所以丰富多彩，是因为有许多物质在黑洞周围绕转，这些物质之间可以发生相互作用。此外，引力并不是唯一的必须遵守的物理定律：角动量守恒定律也必须成立。将这些定律应用于可能被黑洞吸引的大量物质，会引发显著的可观测现象，被称为类星体的奇异天体就是一个很好的例子。**类星体**是位于星系中心的天体，其核心有一个超大质量黑洞，它对附近的物质有影响，这种影响，使它在整个电磁波谱发出的光甚至比某些星系中所有恒星 [64]

还要亮。我们将在第 8 章中讨论类星体和其他类型的“活动星系”，还有缩小版的微类星体——它们的黑洞要比类星体内的黑洞质量小几个量级。现在，让我们回过头来继续探究黑洞周围的物质。

正如我们所看到的，你没有办法直接观测一个孤立的黑洞，因为它根本不会发光，你只能通过黑洞与其他物质的相互作用来探测它。任何落入黑洞的物质都将获得动能，并且与其他下落物质一起形成旋涡。这个旋涡被称为**湍流**，通过湍流的物质会被加热，而这种加热会使原子电离，发出电磁辐射。因此，黑洞对附近物质的作用会导致黑洞周围发出辐射，而黑洞本身不会发出辐射。

黑洞在太空中不是孤立的、没有相互作用的实体。它们的引力场会将所有物质吸过去，无论是附近的气体还是恒星。由于引力随着距离的缩短而急剧增加，如果恒星不幸与黑洞发生了近距离接触，它们就会被撕裂。图 15 就是一个例子。被吸过去的物质中有一部分将被黑洞完全吞噬或吸积。物质不会只是加速冲向黑洞，并飞快地穿过事件视界。相反，在引力吸引物质并使其靠近黑洞时，会有一些精心设计的“求爱仪式”。人们发现，被吸积的物质会聚集成一个特殊的几何形状——通常是盘状。如果引力场是成球状对称的，黑洞将无法决定气体沉积到哪个平面上形成吸积盘——吸积盘的平面位置将由远离黑洞的气体流的性质决定。但是，如果黑洞具有自转，那么无论在半径较大的地方气体如何流动，物质最终都会沉积到垂直于其自转轴的平面上。哪怕是被吸引的物质有一点旋转，都必须考虑角动量守恒，这点在第 3 章中讨论最终会坍缩成黑洞的物质的转动时提到过。旋转意味着物质在失去能量时将沿着非常圆但实际上是螺旋状的轨道向内运动。在黑洞附近，我们在第 3 章中提到的伦泽–蒂林效应意味着，在半径较小

65

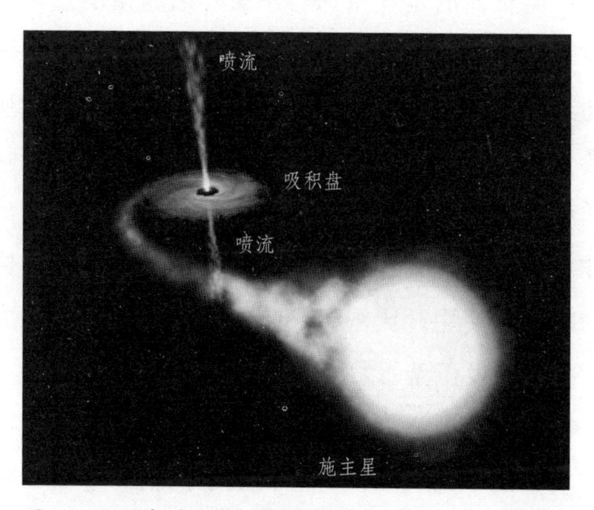

图 15　吸积盘（从中可以看到射出的喷流——见第 8 章）和施主星的效果图。其中施主星正被吸积盘中心黑洞的潮汐力所撕裂

的地方，吸积盘可能会与旋转黑洞的赤道面一致［这个论点中，此效应称为巴丁–彼得森效应（Bardeen–Petterson effect）］。

如果气体是坍缩物质的重要组分，那么气体原子就可以与位于其所在轨道上的其他气体原子发生碰撞，而这些碰撞会导致那些原子中的电子被激发到更高的能级。当这些电子回落到更低的能级时，它们所释放的光子能量恰好是电子所在的较高能级与较低能级间的能量之差。释放出光子就产生了辐射能，这意味着坍缩中的气体云损失了能量。尽管能量被释放了出来，但整体的角动量保持不变。因为角动量依然留存在系统中，所以坍缩中的物质仍然会在某个平面上沿初始净角动量的方向旋转。因此，被吸引的物质将总会形成一个吸积盘——一种可以维持很长时间的物质绕着黑洞运转的结构。由于绕转的物质可以离黑洞非常近，其实际的热度高达1000万度（温度这么高的时候，使用开氏温标还是摄氏温标并没有太大区别），能令吸积盘所发出的辐射包含X射线光子。

66

对牛顿物理学中一些常见方程的简单分析表明，给定质量的下落物质所释放出的引力能，取决于其质量与它旋转落入的黑洞质量的乘积，以及下落物质最终距离黑洞的远近。如图16所示，对于给定质量的，类似黑洞这样产生引力的物体，下落的物质离它越近，释放出的引力势能就越大。可辐射出的能量是下落物质在加速之前位于远处时的能量（使用爱因斯坦著名的公式 $E=mc^2$ 计算，其中 E 是能量，m 是质量，c 是光速）与它在最内稳定圆轨道上的能量之差。

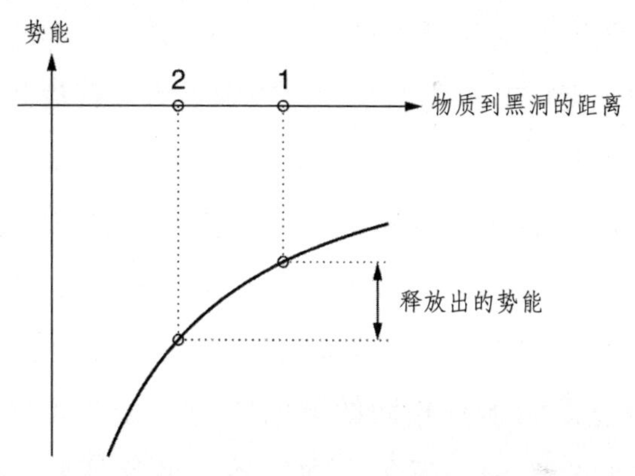

图 16　质量（测试粒子）的势能随着到黑洞距离的减小而减小

尽管聚变是地球未来能源的巨大希望，但它最多只能产生可用能量（$E=mc^2$ 计算出）的 0.7%。相比之下，静质量如果通过电磁辐射或其他辐射从吸积物质中释放出来，能量明显会更多。如第 4 章中所述，吸积物质能够到达离黑洞多近的地方，取决于黑洞的旋转速度。如果黑洞旋转很快，则物质就可

67

以维持在更小（或者说距离黑洞更近的）轨道上绕转的模式。事实上，将物质吸积到旋转的黑洞上，是用质量换取能量的最有效方法。人们认为正是这个过程为类星体提供了燃料。类星体是宇宙中最强大的持续释放能量的场所，我们将在第 8 章中进一步讨论这个问题。

我已经提到过质量和能量是等效的，并且对于史瓦西（无旋转）黑洞来说，原则上可以释放相当于其初始质量 6%的能量。罗伊·克尔找出的爱因斯坦场方程的解表明：旋转黑洞的最内稳定圆轨道的半径比相同质量的无旋转黑洞小得多。原则上可以从克尔黑洞中提取更多的转动能，但前提是下落的物质按照与黑洞本身相同的方向转动。如果物质按照与黑洞自转方向相反的方向转动，也就是说它处于逆行轨道上，那么就只有不到 4%的静止能量会以电磁辐射的形式被释放出来。假如物质坠入一个以最大限度自转的黑洞，并且自转方向与黑洞自转的方向相同，那么原则上如果该物质能够损失足够多的角动量，并且能够在顺行的最内稳定圆轨道上绕转，将有多达 42%的静止能量能以辐射形式被释放出来。

68

它们吃得有多快?

我们在第 6 章提到过，位于人马座 A* 的银河中心黑洞的吸积率是每年亿分之一的太阳质量。听起来似乎不多，但要知道，这相当于每年吞噬 300 个地球。典型类星体的巨大光度所需要的物质落入量，是每年几倍的太阳质量。而我们将在第 8 章中讨论的更小规模的微类星体的典型光度，所需的物质落入量可能是典型类星体的百万分之一。

另一种类似的能量提取过程可能发生在**伽马射线暴**中，通常被称为 GRBs。它是指突然闪烁的强烈的伽马射线束，与遥远星系中的剧烈爆炸有关。20 世纪 60 年代后期，美国卫星首次观测到这些射线束，它们一开始发出的信号被怀疑是来自苏联的核武器。

考虑到物质通过圆盘螺旋落入黑洞的情况是普遍存在的，物理学家们认为对一些重要物理量的大小进行简单而有启发性的计算是非常有帮助的。如果我们考虑的是球面几何而不是圆盘几何，那么某些有趣的限制就会出现。一个特别有说服力的例子来自恒星世界，与吸积盘相比，将它们视为等离子体球要好得多。亚瑟·爱丁顿爵士指出，被激发的电子与恒星热气体中的其他离子碰撞，所释放的辐射将对随后被

69

其拦截的任何物质施加辐射压力。光子可以“散射”（这就意味着“给予能量和动量”）恒星内部被热电离的等离子体中的电子。向外的压力通过静电力（由电荷相互作用产生的与引力类似的力）传递到带正电的离子上：例如氢原子核（也被称为质子）、氦核，还有其他更重的元素的原子核。

以恒星为例，净辐射沿着径向射出，由此产生的辐射压也与将物质向内拉向中心的引力作用方向相反。对于类球形的恒星而言，在向外的辐射压力超过向内的引力并使恒星自己炸开之前，其辐射压有一个最大限制。这个辐射压的最大值被称为**爱丁顿极限**。更高的辐射压力必然来自更高光度的辐射，如果我们知道与物体的距离，就可以根据其亮度估算它的光度。因此，通过一些简化假设——比如将吸积盘视为球体，就可以推断出物体内部的辐射压力大小。这种简单的方法有时被用于粗略估计黑洞的质量：通过观测周围等离子体所发出的辐射光度，并假设这就是达到最大极限值的“爱丁顿光度”（超过这个阈值的光度所施加的辐射压力，会高到足以超过内部质量产生的引力，从而将自身撑爆），就可以估算出它的质量。

在对物质吸积率作出合理假定的情况下，爱丁顿光度可以看成物质所能达到的最大吸积率。这给出了一个被称为爱丁顿比率（在假定的效率下）的最大值。有多种方法可以打破这个最大值的限制，其中之一就是拒绝球对称假设（对于恒星来说还好，但不适用于圆盘几何，我们理解黑洞如何生长就需要用圆盘几何）。 70

如何测量吸积盘内的旋转速度

由于天文学技术的进步，现在，至少在离地球比较近的情况下，可以测量物质绕黑洞运行的速度。最大的挑战之一是：要获得在足够精确的角度范围内的信息非常难。其所需的空间分辨率比通常的光学望远镜高出至少 100 倍，有时甚至是 1000 倍。原则上，用望远镜获得更高分辨率的方法是在更短的波长下观测，或者建造更大的望远镜，尤其注意要减小观测波长与所用望远镜口径的比值。不幸的是，后一种方法非常昂贵，前一种方法则会将通常的可见光观测带到紫外区域，而紫外线是很难穿过地球大气的。与直觉相悖，要实现更小的观测波长与望远镜口径之比，需要在无线电波段（比可见光或紫外线的波长长得多）

观察，因为无线电波可以穿过大气层，但这样一来望远镜的口径几乎要等于地球直径。

这个方法存在的一些技术问题，需要在此稍作讨论。事实证明，多亏法国数学家让·巴普蒂斯·约瑟夫·傅里叶（Jean Baptiste Joseph Fourier）在数学发展71中作出的贡献，即使实际观测区域只是理想状况的稀疏子集，对于望远镜完整孔径所观察到的信号，我们还是能将其中大部分恢复出来。如果将分立天线（每个天线看起来像一个单独的望远镜，请参见图17所示被称为VLBA的甚长基线阵列）的信号相互关联在一起，就可以重构天空中某个区域内的图像，这些图像的精细程度与一个完整地球大小的望远镜所能观测到的图像的精细程度相当。我打个比方来说明这个分辨率有多惊人：假设我站在纽约帝国大厦的顶上，而你在旧金山，在这个距离下，你仍然能看清我的小拇72指指甲（我忽略了地球是一个球体，实际上旧金山和帝国大厦之间并没有直接的视线，但你应该能明白我想说什么）。这意味着使用VLBA，我们可以分辨其他星系中尺度小于一个光月的图像。

同时具有空间意义上的高分辨率和光谱意义上的高分辨率（意味着我们可以非常精确地识别光谱中某个特定的波长），是一种非常强大的结合。哈佛大学

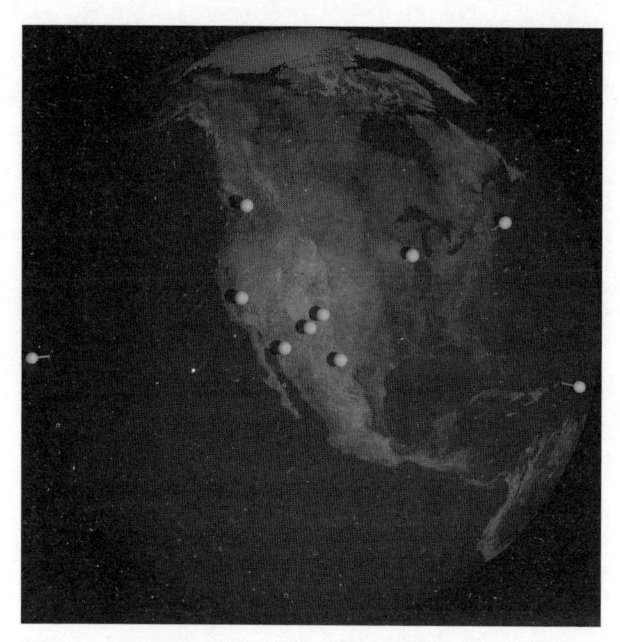

图 17　甚长基线阵列（VLBA）的效果图。这些天线可以合起来提供超高分辨率的照片，相当于地球直径那么大的镜头才能拍出来

的吉姆·莫兰（Jim Moran）所领导的研究小组，利用多普勒效应对附近星系一个名为 NGC 4258 的中心黑洞周围的吸积盘使用 VLBA 进行了观测。他们测量了在整个旋转的吸积盘上波长变化的特定光谱信号（被称为“**水脉泽**”），并利用随着发出脉泽的物质靠近和远离地球时导致的红移和蓝移，来探测物质在黑洞周围给定距离的轨道上运动速度的变化。这些精确的数据证实了物质绕着黑洞转动的轨道正如开普勒定律

所描述的那样，这些轨道如图 18 所示。

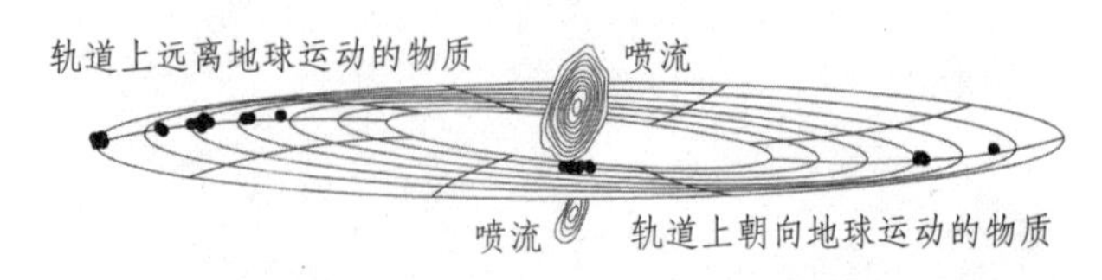

图 18　VLBA 测量了星系 NGC 4258（也被称为梅西耶 106）的吸积盘上绕中心黑洞转动的分立脉泽的分布。这个黑洞的质量是太阳质量的 4000 万倍

旋转的物质

在质量是太阳质量 1 亿倍的黑洞的最内稳定轨道上，物质的角动量是典型星系绕转的角动量的万分之一。显然，要让物质被黑洞所吸积，就需要除去绝大部分的角动量，而这正是通过吸积盘实现的。吸积盘中的轨道可以被近似地看成圆形，尽管实际上它们是两侧略微收缩成螺旋状的。开普勒定律表明，在半径较小的轨道上的物质将比在半径稍大的轨道上的物质运动得更快。这种较差转动[1]使得黑洞能够吸收构成吸

73

1 这里指吸积盘上不同半径处角速度互不相同的现象。

积盘的等离子体：快速旋转的内缘轨道上的物质，会与半径稍大的相邻轨道上旋转较慢的物质发生摩擦，从而产生热量。这种速度上的差异意味着，由于黏性湍流效应，稍大轨道上的物质将被拖快一些；相应地，更靠内缘轨道上的物质将被拖慢一些。因此，由于轨道运动进一步增加，角动量会从内部物质传递到外部物质，同时将物质加热。

总体而言，角动量是守恒的，内部物质可以逐步失去角动量，因而更容易被黑洞吞噬。请注意，如果轨道上的一团物质角动量太大，那么它将远离所绕转的质心——它将因移动得太快而无法靠近。什么样的黏性效应可能与吸积盘内的等离子体有关呢？在这种情况下，原子间的黏度会很小，构成吸积盘的气态等离子体稠度与糖浆相差甚远。实际上，磁场对于将角动量转移出来可能非常重要。磁场从何而来？吸积盘中的等离子体非常热，因此原子被部分电离为电子和带正电的核子。如同詹姆斯·克拉克·麦克斯韦（James Clerk Maxwell）的方程所描述的那样，带电粒子流和移动的电荷会产生磁场。只要存在非常微弱的磁场，它们就可以被较差转动拉伸和放大，并被等离子体的湍流所修正，直至达到所需的黏度。这就是所谓“磁旋转不稳定性”的基础。20 世纪 90 年代

74

初，在弗吉尼亚大学工作的史蒂夫·拜尔巴斯（Steve Balbus）和约翰·霍利（John Hawley）最早意识到了这种机制的重要性。

通过黏性湍流和其他可能的方式，等离子体最终会失去角动量，并在更靠近黑洞的、比其半径更小的轨道上绕转。一旦气态等离子体到达最内稳定轨道，不再需要任何摩擦力就可以落入黑洞，此后就再也看不到它了，但它会增加黑洞的质量和自旋。

吸积盘看起来是什么样的？它们有多热？

我们已经看到，黏性和湍流效应在去除轨道物质的角动量方面起着重要作用，因为它们，物质大可以在更近的地方绕黑洞运动，并被黑洞吞噬。不过，黏性作用导致的一个后果是，整体的轨道螺旋运动被转换为随机的热运动，物质变热了。物质的随机热运动越剧烈，其所拥有的热能就越多，温度也越高。如第 5 章所述，有热量的地方就会有热电磁辐射。每个物体都会发出热辐射，除非它处于绝对零度。

这个加热过程是我们能从吸积盘上观测到高光度辐射的原因。对于类星体中心的超大质量黑洞周围环绕的吸积盘来说，其特征尺度有10亿千米，并且这些吸积盘发出的辐射在光谱上主要分布在可见光和紫外区域。对于微类星体（将在第8章中进行讨论）中质量更小的黑洞周围的吸积盘来说，其大小要比类星体小100万倍，并且辐射以X射线为主。黑洞质量越大，最内层的稳定圆轨道就越大，因此周围的吸积盘也就越冷。

75

质量是太阳100倍的超大质量黑洞周围的吸积盘，最高温度可达100万开尔文；而恒星质量黑洞周围的吸积盘的最高温度，比这还要高100倍。

如何测量黑洞自旋有多快

实际上，你无法直接看到黑洞，因此你也看不到它们在旋转。但仍然有两种主流的方法可以测量黑洞的自旋有多快。如第4章所述，当黑洞自旋得非常快时，黑洞周围稳定轨道上的物质就可能比没有自旋的情况下离黑洞更近。事实证明，离黑洞非常近的轨道上的物质，螺旋下落时会由于强烈的湍流和黏性效应被加热，巨大的热量使它辐射出X射线，这种辐射同时取决于物质被黑洞吞噬前与黑洞有多近。广义相对论预言，谱线形状呈现的某种特征是受辐射物质与黑洞的距离影响。这种特征来自于物质中铁原子的荧光辐射，这一从 X 射线光中提取信息的方法由剑桥大学的安德鲁·法比安（Andrew Fabian）率先提出。

这种测量非常具有挑战性，因为存在许多不可控的因素，比如吸积盘相对于地球的倾斜度，以及实际上来自吸积盘表面的风和外流物质的性质。在吸积盘内缘的附近（沿着我们的视线方向）有着可以揭示黑洞信息的特征，黑洞信息无法通过其他方式被看到。测量恒星质量黑洞自转的其他方法包括测量较大范围的 X 射线谱，用于解释吸积盘的内部区域（更热）和较远区域（逐渐变冷）的不同温度。我们可以根据 X 射线光谱的形状得到吸积盘的倾角，并根据最高温度（假设你知道黑洞的质量及其与地球的距离）得到最内缘的物质在离黑洞多远的地方绕转。杜伦大学的克里斯汀·多恩（Christine Done）正在开发一种类似的方法，以便测量类星体中心的超大质量黑洞的自旋。物质能够在多近的轨道（在被黑洞吞噬之前）上绕转，会告诉你黑洞的自旋有多快。

76

狼吞虎咽的黑洞

事实证明，只有一小部分（估计有 10%，尽管实际上可能比这要高得多）被吸向黑洞的物质能到达事件视界并被吞噬。第 8 章将讨论那些落向黑洞却没有被吞噬到事件视界内的物质发生了什么。在穿过吸积盘时，物质可以像风一样被吹走；而从吸积盘的最内缘半径里会喷出速度非常接近光速的等离子体喷流。如第 8 章所述，没有被黑洞吞噬的东西会旋转而出，形成相当壮观的喷流。

08

黑洞和副产品

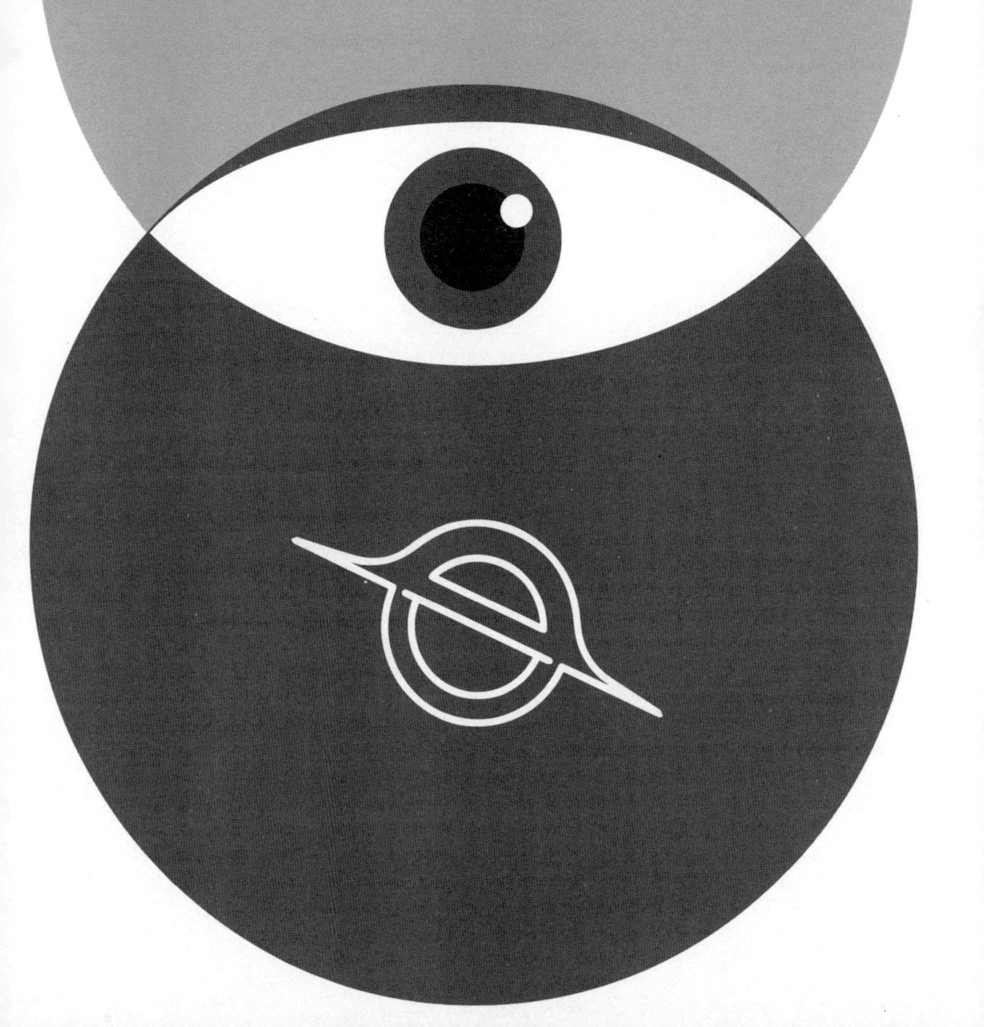

黑洞不只是吸收

77

如果我们的眼睛可以在射电波段或者 X 射线波段观察天空，就会看到一些星系被巨大的气球或等离子体波瓣包裹。这些等离子体中含有运动速度接近光速的带电粒子，发出一定波长范围内的辐射。其中一些星系（比如“活动星系”）所表现出的等离子波瓣是由喷流产生的，它们是从黑洞事件视界周围喷出来的，运动速度快到可以与光速相提并论。罗杰 · 彭罗斯概括性地指出：理论上，从黑洞的能层中提取自转能量是可能的。罗杰 · 布兰福德（Roger Blandford）和罗曼 · 扎纳克（Roman Znajek）明确提出了将旋转黑洞中存储的能量转移到电场和磁场中的方法，从而为产生相对论性等离子体喷流提供动力。从黑洞附近发出喷流的机制也有其他解释，这些解释中哪个才是正确的，正是当前活跃且令人兴奋的研究主题。

最终无论哪种机制被证实，这些喷流都是从黑洞附近（当然是在事件视界外）喷出的高度聚焦且准

78 直的射流。实际上，星系之间的区域并不是真空。与此相反，其中弥散着非常稀薄的被称为**星际介质**的气体。当喷流撞击星际介质时会形成激波，其内部会发生壮观的粒子加速，而被黑洞附近喷流所激发的等离子体，其内部也会发生极其剧烈的运动，从而流出当前的激流区域。随着等离子体膨胀，它会向星际介质传输大量能量。这些等离子喷流中，有许多会延伸至数百万光年外。因此，黑洞对超过事件视界很多光年的宇宙依然有着巨大影响。在本章中，我将描述黑洞对其周围环境的影响以及与周围环境的相互作用。

如第 6 章所述，在大多数星系的中心（可能）有一个黑洞，物质会被其吸积，从而产生电磁辐射。这样的星系被称为活动星系。它们其中一些星系的吸积过程非常强烈，产生的辐射光度极高。这样的星系被称为类星体（这个词源于它们最初被识别为“类似恒星的射电源”，因为它们是遥远的亮度很高的射电点光源）。我们现在知道，类星体是宇宙中已知的最强大的持续能量释放场所。类星体辐射的能量跨越整个电磁频谱，从长波射电波到光学（可见光）波段，再到 X 射线并继续向后。上面提到的射电瓣之所以特别引人注目，是因为它们跨越了数十万光年

（图 19）。射电波段辐射的能量来自那些巨大的波瓣——也就是存有超热的磁化等离子体的地方，其能量是由空间中长距离传输能量的喷流所提供的。高能电子（此处高能是指其传播速度非常接近光速）在行经等离子体波瓣时会受到来自遍布其中的磁场所施加的垂直于运动方向的力。

这种加速使它们发出被称为同步辐射的光子（可能是射电的，也可能是红外的；或者在极短的波长下高能的情况，也可能是 X 射线）。 79

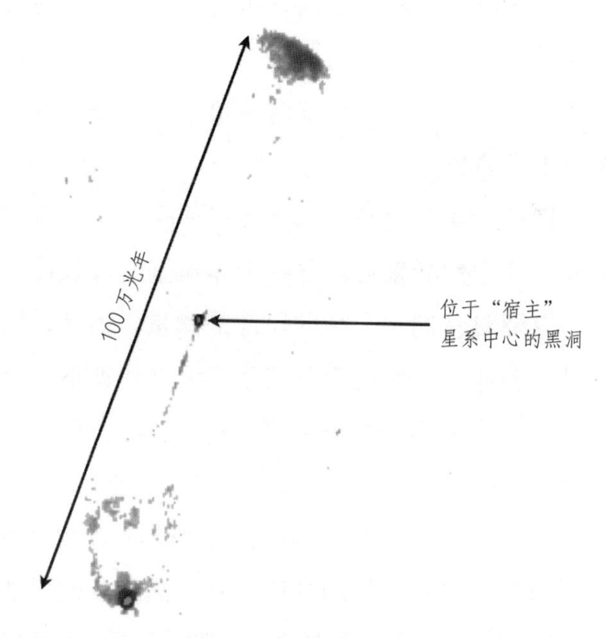

图 19　一个巨大类星体的射电图像。它的范围超过 100 万光年

要了解类星体产生的功率的规模，我们来考虑以下几个典型情况。我用于工作的LED灯，输出功率是10瓦。它们由输出功率高达数十亿瓦的本地发电站提供的电能所点亮（10亿瓦等于10^9瓦或1吉瓦）。太阳的输出功率约为4×10^{26}瓦，是这个发电站功率的10亿亿倍。我们所在的银河系有1000多亿颗恒星，其输出功率接近10^{37}瓦。但是类星体产生的功率甚至可以比银河系的输出功率高100倍以上。请记住，这个功率不是由一个星系或1000亿颗恒星发出的，而是由单个黑洞周围的能量所产生的。这样的辐射可能会对地球上生物的健康造成极大损害，因此可以说我们非常幸运，因为在距离银河系很近的地方没有这样强大的类星体！

人们认为类星体中的喷流可以持续10亿年或更短时间，这个判断来自对这些物体喷流的成长速度的估计，以及对它们长大后的尺寸的测量。因此，用距离、时间和速度之间的简单关系，可以估算整个宇宙中可能观测到的类星体中喷流活动的持续时间。

80

随着这些射电波瓣的扩张，它们的磁场会减弱，波瓣中各个电子的“内部”能量也随之减弱。这两种效应会让辐射强度随着时间的推移和到黑洞距离的增加而减小，强度下降的幅度取决于其中高能电子与低能电子的相对数量。同步辐射的一个特性是，磁场强度越弱，产生射电望远镜可接收长波的辐射所需的电子能量就越高。所以当等离子体波瓣扩展到外部空间时，同步辐射也随之减弱。电子不仅会随着等离子体的膨胀而损失能量，而且随着磁场强度的减弱，只有那些和高能量电子相关的现象才会与望远镜观测到的有关，并且通常情况下，这些电子的数量要远少于低能电子。就类星体的射电波瓣而言，光可以在很短时间内就熄灭。

不过，这场表演还远没有结束，只是奇观转移到了另一个波段。一些非比寻常的事情正在发生：波瓣会在 X 射线波段上亮起来。这是通过被称为“**逆康普顿散射**”的过程发生的。在足够大的磁场中，电子会发出同步辐射，从而失去能量。而我们在此处讨论的另一种能量损失机制，则是通过这些电子与构成宇宙微波背景辐射（CMB）的光子相互作用发生的，这些光子源于大爆炸残留的辐射，宇宙目前正沐浴在这种凉爽的微波辉光中。这些电子可能会与 CMB 中的光子发生碰撞，光子从中获得比碰撞前高得多的能量，电子则相反（要记住整体的能量是守恒的）。特别令人感兴趣的是，当快速运动的电子能量减少到静止电子能量的 1000 倍时（此前是静止电子能量的成百上千倍），恰好可以将 CMB 光子散射为 X 射线光子。

81

高能电子与低能光子通过相互作用产生高能光子。这在某种程度上类似斯诺克中的情况，白色母球（想象这是一个电子）与某个红色的斯诺克球发生碰撞（为了便于说明，请忽略球并没有以光速运动），而红色的球从母球那里获得了大量能量。尽管（希望如此）红球最终会落在台球桌上的一个球袋中，但光子（原本的波长约为 1 毫米）获得的能量是碰撞前的 100 万倍，因此它的波长也缩短了 100 万倍。

美国国家航空航天局（NASA）于 1999 年发射的钱德拉卫星对 X 射线波段很敏感，并且能够在 X 射线波段探测到一对哑铃形的波瓣，就像射电望远镜可以在厘米波段探测到这些双瓣结构一样。图 20 和图 21 显示了在射电波段观察到的双瓣结构的等高线图，以及在 X 射线波段探测到的双瓣结构的灰度图。

实际上，如果我们能够监测这些类星体在整个演化阶段的生命周期（这与生物学家观察青蛙的生命周期类似，从蛙卵到蝌蚪，再到带有很小的腿的蝌蚪，到尾巴粗短的小青蛙，最后到大青蛙和死青蛙），我们将观察到双重结构辐射从射电波段逐渐变为 X 射线波段。首先，射电结构会逐渐消失，直到无法探测到，然后 X 射线结构也将逐渐消失，直至无法探测到。当然，如果喷流重新开始——例如黑洞获得了更多的燃料，那么喷流将为新的发射射电波的双瓣提供燃料，然后再为发射 X 射线的波瓣提供燃料。如图 20 和图 21 所示，在某些类星体中，我们可以同时看到射电和 X 射线的双重结构，而在另一些类星体中，只能看到射电或 X 射线其中之一（图 22）。在一些不同寻常的情况下，我们看到了 X 射线的双重结构，这与先前的喷流活动相对应，但也有一些角度不同的新射电活动，这是因为反向喷射的喷流方向发生了转动，也就是产生了进动。这种现象的一个例子如图 21 所示。 82

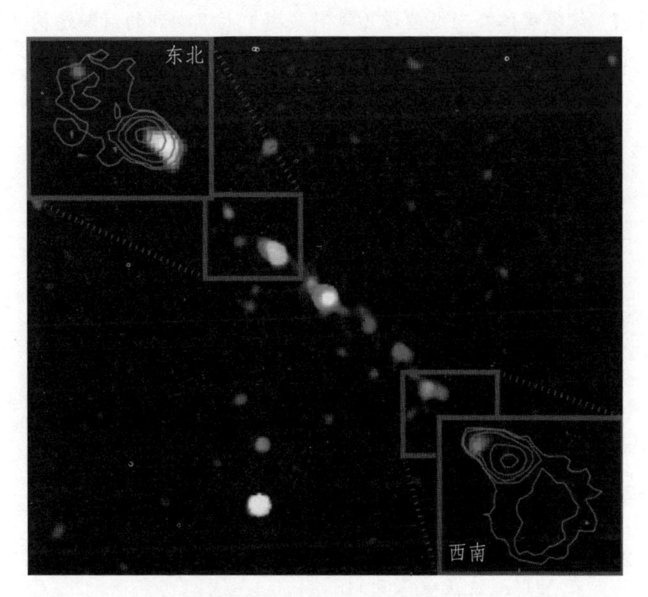

图 20　这个巨大的类星体的范围达到了 50 万光年，并且在射电（以等高线显示）和 X 射线（以灰度显示）波段都具有双瓣结构

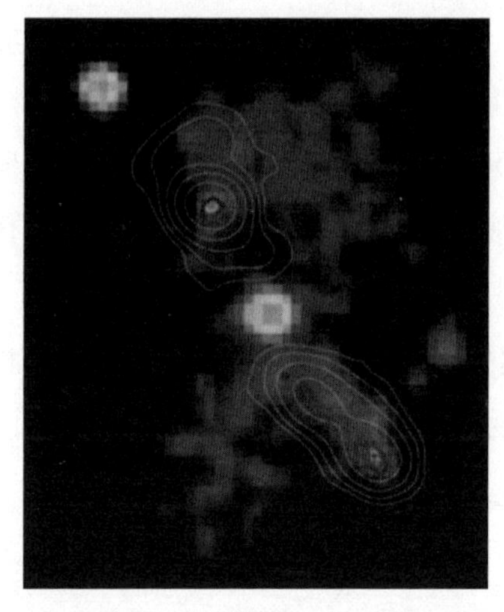

图 21　该类星体在射电波段（等高线图）上观测到的双瓣结构显示出近期的活动与在 X 射线能量（灰度，CMB 光子的逆康普顿散射所揭示的遗迹的辐射）下观测到的双瓣结构的方向不同，这表明喷流轴可能像微类星体的喷流轴一样发生进动

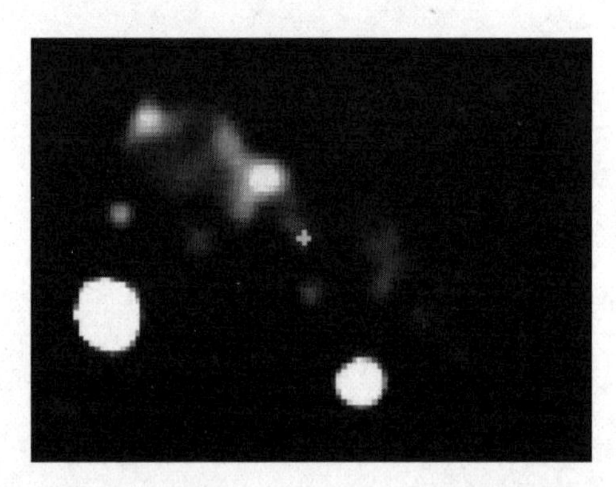

图 22　这张 X 射线图像显示了横跨这个星系的双瓣结构。它只能在 X 射线波段被探测到

许多类星体和射电星系喷流轴的稳定性，揭示了超大质量黑洞自旋的稳定性，这就像陀螺仪一样。为[83]什么某些喷流轴会发生进动而另一些则不会，这个问题的关键在于黑洞喷射点附近影响喷流角动量的因素是什么。究竟是与黑洞本身的自转轴有关，还是与吸积盘内部区域的角动量矢量有关目前尚不清楚，后者[84]由我分别在第 3 章和第 7 章中提到的伦泽–蒂林或巴丁–彼得森效应所决定。我们需要更多数据才能彻底阐明已观测到的现象。但是，有一些来自更靠近我们的较小天体的线索可能表明，喷流轴的进动与吸积盘的角动量有关。

微类星体

到目前为止，我们一直讨论的类星体都是位于活动星系中心的超大质量黑洞。事实证明，还有另一类天体的行为与它们非常相似，但规模小得多。这些质量较小的黑洞可以在离我们更近的地方被观测到。实际上，它们就位于我们自己的银河系中，被称为“微类星体”。尽管大小悬殊，但银河系中的微类星体和其他星系中心的系外类星体一样，都是具有类似物理性质的等离子喷流源。人们认为两者都是由受引力作用落向黑洞的物质所驱动。在微类星体中，黑洞的质量与太阳相当。对于强大的系外类星体而言，其黑洞质量可能比太阳的质量大 1 亿倍。就天体物理学家所关心的事情来说，本地事例的一个重要优势在于其质量较小，因此演化的速度要快得多。它们的演化时标是几天，而不像类星体那样要几百万年。不过，与类星体一样，从所有活动中心附近喷出的喷流都是源于事件视界之外的，而且很可能是从吸积盘的最内缘发出的。

作用于微类星体间的机制非常复杂，而且喷流的发射速度和与其相关的黑洞质量的关系也并不简单。在监测被称为天鹅座 X-3 的黑洞微类星体中的喷流过程时，有时会发现离开黑洞喷流等离子体的速度发生了变化。这是利用延时天文测量法得到的，延时天文测量法就是在一段时间进行连续观测，使我们能够确定等离子体喷流从黑洞附近跑出来时速度有多快。测量结果显示，某一次喷流的速度是光速的 81%，而 4 年后则是 67%。没有迹象表明喷流速度只会随着时间推移而降低，自从发现这个微类星体以来，快速和慢速的喷流都已经被观测到很多次。喷流速度的变化似乎也是银河系中另一个著名微类星体 SS433 的特征，我将在下面对它进行详细介绍。这个微类星体中的喷流速度忽快忽慢，实际上，几天之内它的速度可能是光速的 20%~30%之间的任意值。

对称之美

图 23 显示了银河系中的微类星体 SS433 的射电图像，它距离我们只有 18 000 光年。等离子体喷流的结构投影到我们的天空平面上时，会呈现出醒目的之字形或螺旋形图案。组成喷流的各个等离子体火球，正分别以某个介于光速 20%~30%之间的惊人速度运动。火球运动的方向按照一个固定的周期变化。实际上，喷流的发射轴的进动方式与在皮划艇参考系下看到的运动员划桨方式大致相同，只不过这一过程的时标是 6 个月而不是几秒钟。显然，至少在某些类星体中（图 21）也发生了相同的情况，不过如前文所述，它们的速度慢到我们无法对发生的变化进行恰当的时间采样。

喷流在天空中呈现之字形还是螺旋形直接取决于火球的物理运动方式，以及进行观测的具体时间。喷流的一个显著特征就是它们的对称性：东侧喷流部分

86

的物理运动与西侧喷流部分等大且反向：当一个等离子体火球速度达到光速的28%时，在反向喷流中与之对应的部分速度也是这么大；而对于以22%的光速运动的另一个等离子火球，其反向喷流中与之对应的部分速度也会和它一样。实际上，如果一个喷流看起来具有之字形结构，而另一个喷流看上去则是完全不同的螺旋形结构，这是由于喷流等离子体始终以与光速相当的速度运动，此种情况下会发生相对论性畸变。微类星体的辐射功率相对于系外类星体而言是很小的，但是与太阳微不足道的功率相比仍然非常巨大。太阳的总光度只有 4×10^{26} 瓦，还不到图 23 中的微类星体辐射功率的十万分之一。

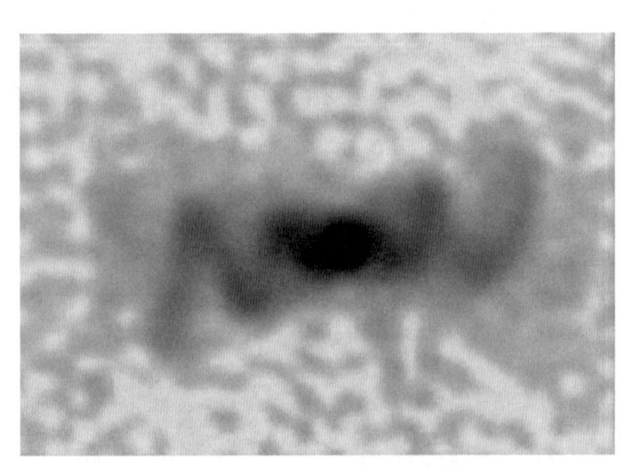

图 23　微类星体 SS433 在射电波段呈现的喷流

喷流的发射

室女座星系团是由 1000 多个星系构成的，距银河系只有 5000 万光年。它的中心是一个被称为 M87［梅西耶 87 的缩写，列在法国天文学家查尔斯·梅西耶（Charles Messier）制作的星表中］的巨大星系。该星系的核心是一个质量是太阳 30 亿倍的超大质量黑洞[1]。从其中发出的是如图 24 所示的非常强的直线喷流。

87

这个喷流在光学波段、射电波段和 X 射线波段都很容易被看到。人们认为，落入物质以每年 2~3 倍太阳质量的吸积率，到达第 6 章中描述的那种吸积盘正在发挥作用的核心区。这个喷流的发射点可能在吸积盘的最内部，其从发射点向外传播的速度非常接近光速，因此我们称之为相对论性喷流。利用我在第 7 章中介绍过的 VLBA 仪器进行连续监测，可知喷流速度非常接近光速，而位于地球大气层之外的哈勃空间望远镜和钱德拉 X 射线卫星，都比其位于地面上时具有更高的灵敏度。在距地球 5000 万光年的位置上，以光速运动的物体每年会在天空中移动 4 毫弧秒。如果我们考虑到一弧只有一度的 1/3600，那么它的四千分

1 根据 2019 年 4 月 10 日事件视界望远镜 M87 中心黑洞成像的测量，该黑洞的质量约为太阳的 65 亿倍。

图 24　从 M87 星系中心的超大质量黑洞，以接近光速喷出的等离子体喷流

88

之一听起来会是一个小到几乎无法测量的角度，但是 VLBA 仪器能很容易分辨这么小的间隔。VLBA 拍摄到了喷流底部的图像，成像尺寸略小于其超大质量黑洞的史瓦西半径的 30 倍。。

图 25 显示了源于 M87 中超大质量黑洞的相对论性喷流的等离子体射电辐射的波瓣和羽流的示例。

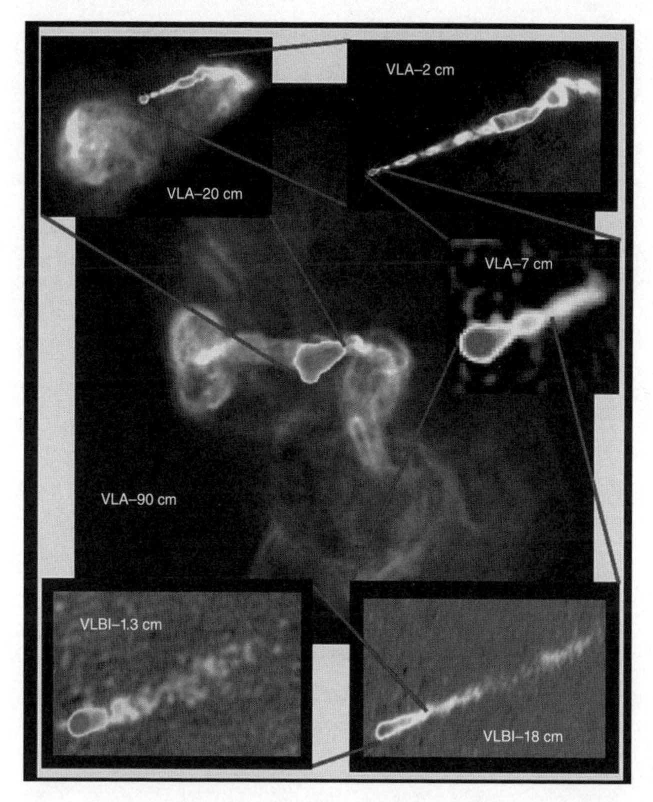

图 25　源于 M87 星系中心的超大质量黑洞所发出的相对论性喷流的射电辐射波瓣和羽流

为了进一步说明膨胀的波瓣与相对论性喷流有关，图 26 展示了一个在天空中延伸了 6 度的示例，并呈现出了用于观测的望远镜阵列，以便让人能够感受其尺度。依拉娜·费恩（Ilana Feain）和她的同事使用的望远镜是澳大利亚望远镜致密阵列。

图 26　月亮和澳大利亚望远镜致密阵列的光学照片与半人马座 A 的无线电图像的合成照片

相对论性喷流从黑洞附近发射的机制目前还只是推测，还不具有普适性。不过，来自世界各地的不同团队进行的各项独立研究中，绝大多数证据表明该理论的基本细节是正确的。除了宽泛的图像以外，这些机制及其详细功能还属于推测，只是在光子不足且具有选择效应的情况下被耐心检验过。证明不属于科学，但证据属于科学。我们之所以受到阻碍，是因为即使当今已部署的最先进的成像技术，也无法区分并识别释放了大部分能量的最小区域，不过，利用功能强大的计算机进行数值模拟，就可以突破当前技术的限制。最新发表的模拟结果表明，吸积盘发出的喷流完全可以由广义相对论效应进行解释。这些模拟将组分和公理作为已知输入，允许喷流和吸积盘演化到其特性可以与最新观测结果相匹配的尺度。

89

那么，我们现在对宇宙中黑洞的质量了解多少呢？看起来它们分为两大类。首先是那些质量与恒星类似的黑洞。这些恒星质量黑洞的质量是太阳质量的3~30倍，它们来自烧光了全部燃料的恒星。

91

然后就是超大质量黑洞，它们能达到约100亿颗太阳质量。正如我们已经讨论过的，它们存在于包括我们自己的银河系在内的星系中心，并且与活跃星系和类星体的种种奇特现象有关。

我们已经讨论过物质掉入黑洞，但当一个黑洞掉入另一个黑洞时会发生什么？这不是一个抽象的问题，因为人们已经知道可能存在双黑洞。在这样的天体中，两个黑洞会互相绕转。人们认为由于发出了引力辐射，双星中的黑洞将失去能量并以螺旋形向内互相绕转。在这种螺旋运动的最后阶段，广义相对论会达到临界点，两个黑洞突然合并为具有常规事件视界的单个黑洞。在一个双星系统中，两个超大质量黑洞合并所产生的能量是惊人的，有可能超过可见宇宙中所有恒星的所有光。它的大部分能量都被注入引力波，这些时空曲率的涟漪会以光速在整个宇宙中传播。对这种波存在证据的搜寻尚在进行。人们设想，当引力波经过像长杆一样的物质时，其波长会在时空曲率的涟漪穿过时随之上下波动。如果可以使用诸如激光干涉之类的技术来测量这些微小的波长变化，就能得到一种可以探测宇宙中其他地方产生的引力波的方法。目前已经建成，以及更多还在计划中的地基和天基引力波探测器，都可能探测到来自黑洞合并的信息。实际上，引力波非常难检测，需要非常强力的能量源才有机会进行此类实验，而在这些强力源的候选名单中，黑洞合并居于首位。在撰写本文时，人们尚未直接检测到引力波，但实验仍在进行[1]。

92

自 1915 年爱因斯坦提出广义相对论以来，我们最好的引力论已接受了无数次考验。事实证明，与被其替代的牛顿经典理论相比，广义相对论的实验具有更好的一致性。但如果要对广义相对论在极限状况下进行检验，那么你可以期待黑洞会成为现代物理学这一

1 2016 年 2 月 11 日激光干涉引力波天文台（LIGO）、处女座干涉仪（Virgo）研究团队共同发布结果：他们于 2015 年 9 月 14 日首次探测到引力波现象。

基石的终极测试地。此种情况下，引力在最小的空间区域中表现得最强，因此量子效应会很重要，而这正是广义相对论可能会崩溃的地方。不过，广义相对论也可能在宇宙中的大尺度上失效。当然，目前最热门的话题是广义相对论在解释宇宙最大尺度上的加速膨胀时的完备性。讨论广义相对论的偏差，可能会与加速膨胀和暗能量有关。如果探测到源自黑洞并合的引力波，或观测结果拓展了我们对发生在这些引人入胜的物体附近的基本物理学的理解，那我们就有机会见证爱因斯坦的理论是能够经受住检验的，还是需要用某些新理论来代替的。

我们为什么研究黑洞

研究黑洞的原因有很多，第一个原因是：它开启了对物理参数空间的探索，即使是国际财团的预算也无法独立胜任这一工作。黑洞系统代表了我们所能探索的极端环境，我们能借此研究极端情况下的物理学。它们将广义相对论和量子力学结合起来，但统一尚未实现，并且仍是物理学的前沿问题。第二个原因是：试图理解黑洞现象引起了科学家和许多有思想的

93

外行人的兴趣，让许多人被科学所吸引，愿意去了解我们周围宇宙的伟大之处。第三个原因也许会令人惊讶，研究黑洞给了尘世一些副产品。对黑洞的研究怎么可能改变我们的生活？答案是这种事情已经发生了。当我将这本小书的最后几句话输入笔记本电脑时，它会同时通过802.11Wi-Fi协议将我的工作备份到我大学的服务器上。这项复杂而巧妙的技术源于在射电波段寻找爆发黑洞的某个特定特征时的研究。这项研究由罗恩·埃克斯（Ron Ekers）所领导的团队完成。他们想要检验马丁·里斯（Martin Rees）（现在是皇家天文学家）提出的模型。在约翰·奥沙利文（John O’Sullivan）的带领下，来自澳大利亚心灵手巧的无线电工程师发明了一种干扰抑制算法，本来是想将它用于探测来自遥远空间的微弱信号这一棘手的工作，但他们随即意识到，这项技术还可以应用于地球上的通信传输。因此，黑洞有能力改写物理学，重新激发我们的想象力，甚至革新我们的技术。黑洞有许多副产品，它们都远远超出了其事件视界。

全书完

名词对照

A

爱丁顿比率	Eddington rate
爱丁顿光度	Eddington luminosity
爱丁顿极限	Eddington limit
爱因斯坦—罗森桥	Einstein–Rosen bridge
暗星	dark star
澳大利亚望远镜致密阵列	Australia Telescope Compact Array

B

巴丁—彼得森效应	Bardeen-Petterson effect
白矮星	white dwarf
白洞	white hole
表面引力	surface gravity

C

参考系	frame of reference
参考系拖曳	frame dragging
测地线	geodesic

超大质量黑洞	supermassive black holes
潮汐力	tidal forces
虫洞	wormhole

D

导星	guide star
等离子体波瓣	plasma lobes
电子简并压力	electron degeneracy pressure
动能	kinetic energy
度规	metric

G

伽马射线暴	gamma-ray bursts
固有时	proper time
惯性坐标系	inertial frame of reference
光速	speed of light
光锥	light cone
光子	photon
光子球	photon sphere
广义相对论	general relativity

H

哈勃空间望远镜	Hubble Space Telescope
海森堡不确定性原理	Heisenberg uncertainty principle
海王星	Neptune
氦	helium
恒星质量黑洞	stellar mass black hole

红移	redshift
皇家天文学会	Royal Astronomical Society
活动星系	active galaxies
霍金辐射	Hawking radiation
霍金温度	Hawking temperature

J

极端克尔黑洞	extreme Kerr black hole
简谐运动	simple harmonic motion
角动量守恒	conservation of angular momentum
较差转动	differential roration
聚变	fusion
绝对零度	absolute zero

K

卡纳维拉尔角	Cape Canavaral
开普勒定律	Kepler's laws
克尔黑洞	Kerr black hole

L

类星体	quasars
零测地线	null geodesic
伦泽—蒂林效应	Lense-Thirring effect
裸奇点	naked singularity

M

闵可夫斯基时空	Minkowski spacetime

N

能层	ergosphere
逆康普顿散射	inverse Compton scattering
逆行轨道	retrograde orbits
牛顿引力常数	Newton's gravitational constant

P

普朗克面积	Planck area

Q

奇点	singularity
钱德拉卫星	Chandra satellite

R

热辐射	thermal radiation
热力学第二定律	second law of thermodynamics
热力学第一定律	first law of thermodynamics
日食	solar eclipse

S

熵	entropy
射电波瓣	radio lobes
甚长基线阵列	Very Long Baseline Array（VLBA）
时间膨胀	time dilation
时空	spacetime
时空曲率	curvature of spacetime

时空图	spacetime diagrams
时序保护猜想	chronology protection conjecture
史瓦西半径	Schwarzschild radius
史瓦西黑洞	Schwarzschild black hole
世界线	worldline
事件视界	event horizon
双星	binary star
水脉泽	water maser
顺行轨道	prograde orbits

T

太阳系	Solar System
逃逸速度	escape velocity
天炉座星系团	Fornax cluster
同步辐射	synchrotron radiation
湍流	turbulence

W

微类星体	microquasars
稳态极限面	static limit

X

吸积盘	accretion disc
狭义相对论	special relativity
相对论性喷流	relativistic jet
信息悖论	information paradox
信息论	information theory

星际介质	intergalactic medium
虚粒子	virtual particles

Y

银道面	Galactic plane
银心	Galactic centre
引力红移	gravitational redshift
引力势阱	gravitational potential well
宇宙监督猜想	cosmic censorship conjecture
宇宙微波背景	Cosmic Microwave Background (CMB)

Z

自适应光学	adaptive optics
真空	vacuum
蒸汽机	steam engine
质点	point mass
质心	centre of mass
中子简并压力	neutron degeneracy pressure
中子星	neutron star
自旋	spin

英文原文

Katherine Blundell

BLACK HOLES

A Very Short Introduction

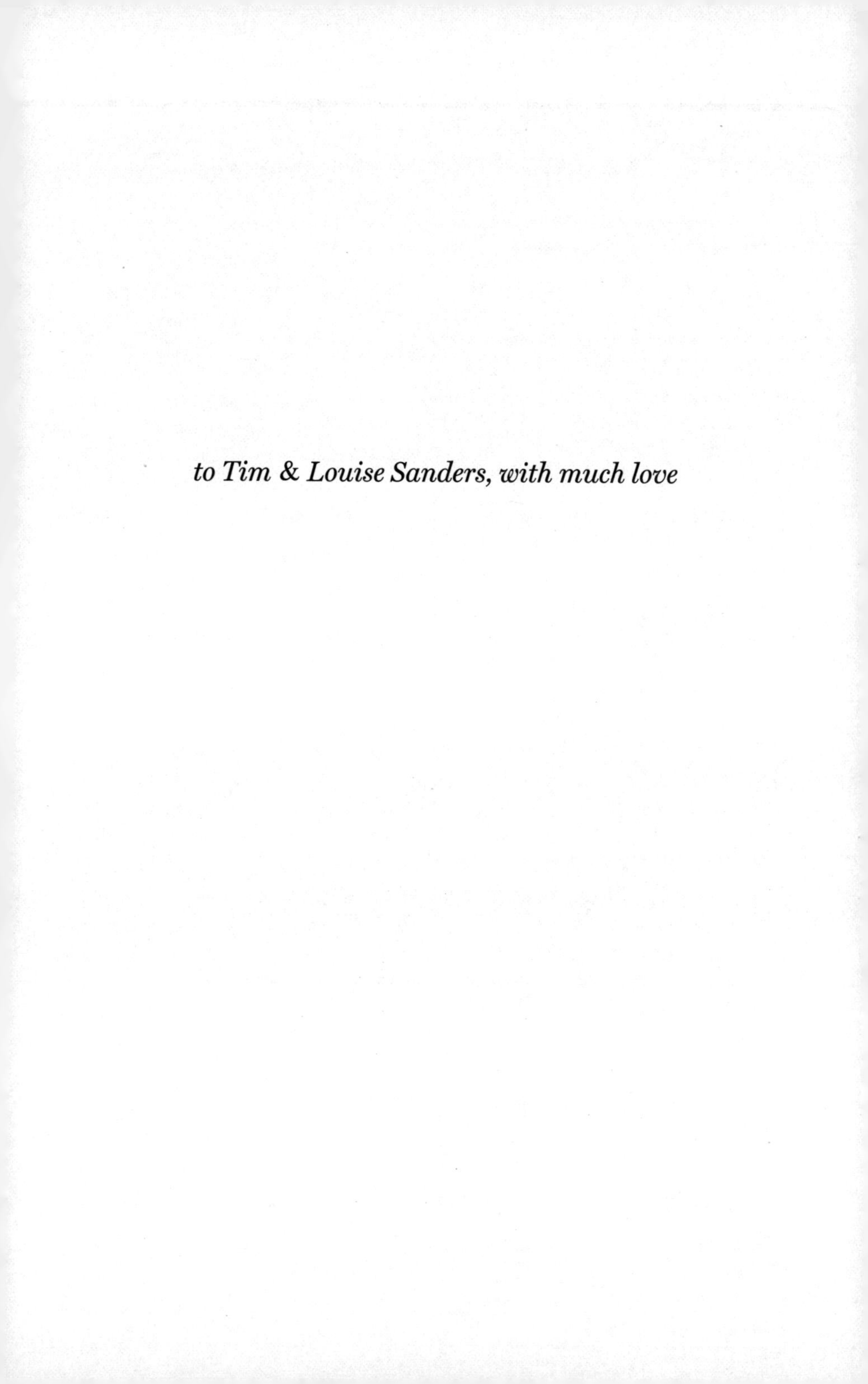

to Tim & Louise Sanders, with much love

Acknowledgements

I should like to record my warm thanks to Phillip Allcock, Russell Allcock, Steven Balbus, Roger Blandford, Stephen Blundell, Stephen Justham, Tom Lancaster, Latha Menon, John Miller, and Paul Tod for many helpful comments on drafts of this book, to Stephen Blundell for preparing the diagrams, and to Steven Lee for assistance with the optical observations.

KMB
Oxford,
April 2015

Contents

List of illustrations

Chapter 1
What is a black hole?

A black hole is a region of space where the force of gravity is so strong that nothing, not even light, can travel fast enough to escape from its interior. Although they were first conceived in the fertile imaginations of theoretical physicists, black holes have now been identified in the Universe in their hundreds and accounted for in their millions. Although invisible, these objects interact with, and can thus influence, their surroundings in a way that can be highly detectable. Exactly what the nature of that interaction is depends on proximity relative to the black hole: too close and there is no escape, but further afield some dramatic and spectacular phenomena will play out.

The term 'black hole' was first mentioned in print in an article by Ann Ewing in 1964, reporting on a symposium held in Texas in 1963, although she never mentioned who coined the expression. In 1967, American physicist John Wheeler needed a shorthand for 'gravitationally completely collapsed star' and began to popularize the term, though the concept of a collapsed star was developed by fellow Americans Robert Oppenheimer and Hartland Snyder back in 1939. In fact, the mathematical foundations of the modern picture of black holes began rather earlier in 1915, with German physicist Karl Schwarzschild solving some important equations of Einstein's (known as the field equations in his General Theory of Relativity) for the case of an isolated non-rotating mass in space.

Two decades later in the UK, a little before Oppenheimer and Snyder's work, Sir Arthur Eddington had worked out some of the relevant mathematics in the context of investigating work by the Indian physicist Subrahmanyan Chandrasekhar on what happens to stars when they die. The physical implications of Eddington's calculations, namely the collapse of massive stars when they have used up all their fuel to form black holes, Eddington himself pronounced to the Royal Astronomical Society in 1935 as being 'absurd'. Despite the apparent absurdity of the notion, black holes are very much part of physical reality throughout our Galaxy and across the Universe. Further advances were made in the United States by David Finkelstein in 1958, who established the existence of a one-way surface surrounding a black hole whose significance for what we shall study in the coming chapters is immense. The existence of this surface doesn't allow light itself to break free from the powerful gravitational attraction within and is the reason why a black hole is black. To begin to understand how this behaviour might arise we need to first understand a profound feature of the physical world: there is a maximum speed at which any particle or any object can travel.

How fast is fast?

A law of the jungle is that if you want to escape a predator you need to run fast. Unless you have exceptional cunning or camouflage, you will only survive if you are swift. The maximum speed with which a mammal can escape an unpleasant situation depends on complex biochemical relationships between mass, muscle strength, and metabolism. The maximum speed with which the most rapidly travelling entity in the Universe can travel is that exhibited by particles that have no mass at all, such as particles of light (known as *photons*). This maximum speed can be given very precisely as 299,792,458 metres per second, equivalent to 186,282 miles per second, which is almost approaching a million times faster than the speed of sound in air. If I could travel at the speed of light, I would be able to travel from my home in the UK to Australia in one fourteenth of a second, barely time to blink. Light travelling from our nearest star, the Sun, takes just eight minutes to travel to us. From our outermost planet, Neptune, it's a journey time of just a few hours for a photon. We say that the Sun is eight light-minutes away from Earth and that Neptune is a few light-hours away from us. This has the interesting consequence that if the Sun stopped shining or if Neptune suddenly turned purple, no one on Earth could find out about such important information for eight minutes or a few hours respectively.

Let's now consider how fast light can travel from even more immensely distant points in space back to Earth. The Milky Way, the Galaxy in which our Solar System resides, is a few hundred thousand *light-years* across. This means that light takes a few hundred thousand years to travel from one side of the Galaxy to the other. The Fornax cluster is the nearest cluster of galaxies to the local group of galaxies (of which the Milky Way is a significant member) and is hundreds of millions of light-years away from us. Thus, an observer on a planet orbiting a star in a galaxy within the Fornax cluster looking back to Earth right now might, if equipped with appropriate instrumentation, see dinosaurs lumbering around on Earth. However, it is only the mind-boggling vastness of the Universe that makes the motion of light look sluggish and time-consuming. The role of the speed of light as a mandatory upper limit has an intriguing effect when we start to consider how to launch rockets into space.

Escape velocity

If we wish to launch a rocket into space but its launch speed is too slow then the rocket will have insufficient *kinetic energy* to break free from the Earth's gravitational field. However, if the rocket has just enough speed to escape the gravitational pull of the Earth, we say it has reached its *escape velocity*. The escape velocity of a rocket from a massive object such as a planet is larger the more massive the planet is and larger the closer the rocket is to the **centre of mass** of the planet. The escape velocity v_{esc} is written as $v_{esc} = \sqrt{2GM/R}$ where M is the mass of the planet and R is the separation of the rocket from the planet's centre of mass and G is a constant of Nature known as Newton's gravitational constant. Gravity always acts so that it pulls the rocket towards the centre of the planet or star in question, towards a point known as the **centre of mass**. However, the value of the escape velocity is completely independent of the mass of the rocket. Thus, the escape velocity of a rocket at Cape Canaveral, some 6,400 km away from the centre of mass of Planet Earth, takes the same value, just over 11 km/s or approximately 34 times the speed of sound (which may be written as Mach 34), irrespective of whether its internal payload is a few feathers or several grand pianos. Now, suppose we could shrink the entire mass of Planet Earth so that it occupies a much smaller volume. Let's say that its radius becomes one quarter of its current value. If the rocket was launched at a distance of 6,400 km away from the centre of mass, its escape velocity would remain the same. However, if it relocated to the new surface of the shrunken Earth 1,600 km from its centre, then the escape velocity would be double the original value.

Now suppose some disaster occurs with the result that the entire mass of the Earth were shrunk to a point, having no spatial extent whatsoever. We call such an object a *singularity*. It has now become a 'point mass', a massive object that occupies zero volume of space. At a very small distance of only one metre away from this singularity, the escape velocity would be much larger than it was at 1,600 km (and in fact would be about 10% of the speed of light). Closer to the singularity still, just under one centimetre away the escape velocity would be equal to the speed of light. At this distance, light itself would not have sufficient speed to escape this gravitational pull. This is the key idea to understand how black holes work.

It is worth clarifying use of the word 'singularity'. We do not believe that at the end point of a continuing gravitational collapse the matter goes down to a geometric point but rather that our classical theory of gravity breaks down and we enter a quantum regime. From here on, we will use the term singularity to refer to this ultra-dense state.

The event horizon

Now imagine you are an astronaut flying a spacecraft and that you are approaching this singularity. While still at some distance from it, you could always throw your engines into reverse and retreat from it. But the closer you get, the harder a dignified retreat becomes. Eventually you reach a distance from which it is impossible to escape, no matter how powerful your onboard engines are. This is because you have reached the *event horizon*, a mathematically-defined spherical surface, which is defined as being the boundary inside of which the escape velocity would exceed the speed of light. For our thought-experiment about Earth collapsed to a point, this surface would be a sphere of radius only one centimetre with the singularity at its centre, easy enough perhaps for our spacecraft to avoid. However, the event horizon becomes much larger when the black hole is formed from a collapsed star rather than a collapsed planet. The event horizon has an important physical consequence: if you are on that surface or inside it, the laws of physics simply won't allow you to escape because to do so you would need to break the universal speed limit. The event horizon is a mandatory level of demarcation: outside it you have freedom to determine your destiny; inside it, and your future remains unalterably locked within.

The radius of this spherical surface is named in honour of Karl Schwarzschild, who was mentioned earlier. While a soldier in World War I, Schwarzschild provided the first exact solutions of Einstein's famous field equations that underpin general relativity. The Schwarzschild radius is written as $R_S = 2GM/c^2$ where M is the mass of the black hole, G is Newton's gravitational constant, and c is the speed of light. Using this formula, the Schwarzschild radius of the Earth comes out to be just under one centimetre. Similarly, the Schwarzschild radius of the Sun is found to be 3 km, meaning that if the mass of our Sun could all be squashed into a singularity, then at just 3 km away from this point the escape velocity would be equal to the speed of light. A black hole one billion times more massive than the Sun (i.e. having a mass of 10^9 solar masses) would have a Schwarzschild radius one billion times larger (the Schwarzschild radius of a point mass that is not rotating simply scales directly with its mass). As I describe in Chapter 6, such mammoth black holes are believed to be at the centres of many galaxies.

This description of the event horizon can be reasonably thought of within Newtonian physics. Indeed, physical entities resembling black holes were imagined centuries before Einstein and others profoundly changed our understanding of space and time. The principal thinkers who imagined 'dark stars' that resemble black holes were John Michell and Pierre-Simon Laplace, starting back in the 18th century, and I will now explain what they did.

One of the remarkable things about astronomy is how much you can discover about the Universe even when you are stuck on planet Earth. For example, no human being has ever visited the Sun, and yet the presence of helium in the Sun was detected in the late 19th century by analysing the spectrum of sunlight. This is particularly remarkable as this constituted the discovery of the element helium itself; it was found on the Sun long before being detected on Earth. Even earlier, in the 18th century, some of the ideas behind black holes were beginning to be formulated, and in particular the idea of what is called a dark star. The person who made the first imaginative leap was very much a product of his time.

John Michell

The Georgian era was, in England, a time of relative peace. The English Civil War was long in the past, and England had become a land of relative domestic tranquillity (the rise of Napoleonic France was still some way off). Like his father before him, the Reverend John Michell (Figure 1) received a university education and entered the Church of England. As a rector in Thornhill, West Yorkshire, Michell was able to continue his scientific research, following up his interests in geology, magnetism, gravity, light, and astronomy. In common with other scientists working in England at the time, such as the astronomer William Herschel and the physicist Henry Cavendish (who was a personal friend), Michell was able to ride the wave of the new Newtonian thinking. Sir Isaac Newton had revolutionized the way in which the Universe was perceived, formulating his law of gravitation which explained the orbits of the planets in the Solar System as being due to the same force that caused his famous apple to drop from the tree.

1. John Michell, polymath.

Newtonian ideas allowed the Universe to be studied using mathematics, and this fresh breed of scientists was able to deploy this novel world-view into different fields. Michell was particularly concerned to use Newtonian thinking to estimate the distance to nearby stars by using measurements of the light they emitted. He came up with various schemes to do this, by relating a star's brightness to its colour; he also considered *binary stars* (pairs of stars gravitationally bound to one another) and how their orbital motions could give useful dynamical information. Michell also investigated how stars tend to cluster in particular areas of the sky, testing this against a random distribution and inferring gravitational clustering. None of these ideas was practicable at the time: few binary stars were known (though Herschel was producing some impressive catalogues of various double stars and new objects) and the relationship between a star's brightness and its colour turned out to be not quite as Michell had thought it was. Nevertheless, Michell was straining to do for the wider Universe what Newton had done for the Solar System: allow a scientific, rational, and dynamical analysis of observations to provide new information about the properties, masses, and distances of the heavenly bodies.

One particular insight that came to Michell followed from the idea that particles of light are, in Michell's words, 'attracted in the same manner as all other bodies with which we are acquainted; that is, by forces bearing the same proportion to their vis inertiae [by which he meant mass], of which there can be no reasonable doubt, gravitation being, as far as we know, or have any reason to believe, an universal law of nature.' Such particles emitted from a large star would, he reasoned, be slowed down by the gravitational attraction of the star. Thus the starlight reaching Earth would be slower. Newton had shown that light slows down in glass, and this explained the principle of refraction. If starlight was indeed similarly slowed, Michell argued that it might be possible to detect this slowing by examining starlight through a prism. The experiment was tried, not by Michell, but by the Astronomer

Royal, the Reverend Dr Nevil Maskelyne, who looked for the diminishing of the refractability of starlight. Cavendish wrote to Michell to tell him that it hadn't worked and that 'there is not much likelyhood [sic] of finding any stars whose light is sensibly diminished'. Michell was dismayed, but such astronomical speculations required much guessing of imponderables: was starlight affected by the gravitational attraction of the star from which it is emitted? Michell couldn't be sure. But he was bold enough to make an interesting prediction.

If a star was sufficiently massive, and gravity really did affect starlight, then the gravitational force could be sufficient to hold back the particles of light completely and prevent them from leaving. Such an object would be a *dark star*. This little-known cleric writing in his rectory in Yorkshire had thus been the first person to conceive of a black hole. However, so far Michell's own programme of measuring the distances to stars lay in tatters. What was more, his health had been indifferent and this had stopped him using his telescope. Cavendish wrote to him a consoling letter: 'if your health does not allow you to go on with [the telescope] I hope it may at least permit the easier and less laborious employment of weighing the world.' This singular example of a joke from Cavendish (who was notoriously buttoned up) refers to another idea that Michell had conceived. 'Weighing the world' meant an experiment in which two large lead spheres at either end of the beam of a torsion balance are attracted by two stationary lead spheres. This allows one to measure the strength of the gravitational force, and thereby infer the weight of the Earth. No one had ever done this before. Michell's idea was brilliant, but he didn't live to complete the project. Instead, Michell's experiment was performed by Cavendish and is now known as Cavendish's experiment. This transfer of credit to Cavendish is more than compensated for by the numerous breakthroughs made by Cavendish which he neglected to publish and were later attributed to subsequent researchers (including 'Ohm's' law and 'Coulomb's' law).

Pierre-Simon Laplace

On the other side of the English Channel, Pierre-Simon Laplace did not enjoy the tranquil idyll afforded by the peaceful period of the English Enlightenment. Laplace lived through the French Revolution, though his career prospered as he influenced the newly founded Institut de France and the École Polytechnique. He even spent a period as Minister for the Interior under Napoleon, a short-lived appointment the Emperor came to regret. Napoleon realized that Laplace was a first-rate mathematician but as an administrator he was worse than average. Napoleon later wrote of Laplace that 'he sought subtleties everywhere, conceived only problems, and finally carried the spirit of "infinitesimals" into the administration'. Napoleon had other administrators to call upon, but the world has had few mathematicians as productive and insightful as Laplace. He made pivotal contributions to geometry, probability, mathematics, celestial mechanics, astronomy, and physics. He worked on topics as diverse as capillary action, comets, inductive reasoning, solar system stability, the speed of sound, differential equations, and spherical harmonics. One of the ideas he considered was dark stars.

In 1796 Laplace published his *Exposition du système du monde*. Written for an educated public, this book describes the physical principles on which astronomy is based, the law of gravity and the motion of the planets in the Solar System, and the laws of motion and mechanics. These ideas are applied to various phenomena, including the tides and the precession of the equinoxes, and the book also contains Laplace's speculations on the origin of the Solar System. One particular passage is of special relevance to our story. Laplace made a calculation of how large an Earth-like body would need to be so that its escape velocity was equal to that of light. He showed, quite correctly, that the gravitational strength on the surface of a star, with density comparable to that of Earth but with a diameter of about 250 times that of the Sun, would be so intense that not even light would be able to escape. Thus, he reasoned, the largest bodies in the Universe would therefore be invisible. Could they still be lurking, undetectable in the dark night sky, while we imagined that the only things 'out there' were the bright luminous objects that we can see? The Hungarian astronomer Franz Xaver von Zach requested that Laplace provide the calculations that led to this conclusion, and Laplace obliged, writing this up (in German) for one of the journals that von Zach edited.

However, Laplace was becoming aware of the wave theory of light. Both Michell's and Laplace's ideas were based in part on the corpuscular theory of light. If light were to consist of tiny particles, then it seemed reasonable that these particles would be affected by a gravitational field and would be bound forever to a star of sufficient size. But the early 19th century saw a number of experiments which seemed to give greater credence to the wave theory of light. If light were instead a wave, then it was harder to see that it should be affected by gravity. Laplace's dark star prediction was quietly omitted from later editions of *Exposition du système du monde*. After all, Michell and Laplace had been conjecturing and exploring theory, rather than being driven by the need to explain observations and thus, this idea was forgotten for a while. The objects imagined by Michell and Laplace were thus 'dark stars', enormous objects in the Universe which by virtue of their mass could sustain planetary systems but by virtue of this same overwhelming bulk could not be observed via the radiation of light. Starlight emitted from the surfaces of Michell's and Laplace's dark stars would be too sluggish to overcome the intense surface gravity. What Michell and Laplace could not have guessed was that such gargantuan accumulations of mass would be unstable to collapse. Moreover, in their collapse they would puncture the very fabric of space and time and give rise to a singularity. Thus 'black holes' are not 'dark stars' and to take the argument forward and begin to meet up with the astronomical discovery of black holes we will first need to understand the nature of spacetime.

Spacetime

Our everyday experience leaves us comfortable with the notion that the tangible Universe may be described by one temporal (or time) coordinate t and three spatial coordinates (for example x, y, and z along three mutually perpendicular axes, a construct invented by René Descartes and known as Cartesian coordinates). In 1905, Einstein published his revolutionary paper on Special Relativity, the relativity of motion and stationarity. In 1907, Hermann Minkowski showed how these results could be understood more deeply by considering a four-dimensional spacetime whose points, specified now by the 4-D coordinate (t, x, y, z), correspond to 'events'. An event is something that happens at a particular time (t) and at a particular place (x, y, z). Such 4-D coordinates in what is known as Minkowski spacetime specify exactly where and when an event occurs. Einstein's special theory of relativity could be formulated in terms of Minkowski's spacetime and provides a convenient description of physical processes in different frames of reference that move relative to one another. A 'frame of reference' is simply the perspective possessed by a particular observer. Einstein called this theory 'special' because it deals only with a particular case, namely reference frames that are non-accelerating (called inertial frames of reference). The special theory can only be applied to uniformly moving, non-accelerated, frames of reference. If you drop a stone, it accelerates towards the ground. The frame of reference attached to the stone is an accelerating frame of reference and cannot be treated by Einstein's special theory. Where you have gravity, you have acceleration.

This drawback prompted Einstein to formulate a *general* theory of relativity, which he published a decade after his special theory. What he found was that whereas Cartesian space and Minkowski spacetime were rigid frameworks in which objects 'live, move and have their being', spacetime was actually a more responsive entity: it could be curved and otherwise deformed by the presence of mass. Once mass is present in a physical situation then the following inextricably linked behaviour describes reality, neatly summarized by John Wheeler:

- mass acts on spacetime, telling it how to curve
- spacetime acts on mass, telling it how to move

This behaviour is quantified by Einstein's field equations within General Relativity, which relate the curvature of spacetime to the gravitational field.

Physicists talk about a *gravitational potential well* as surrounding a massive object. The cartoon shown in Figure 2 encapsulates how the spacetime is distorted in the vicinity of a couple of black holes, where each region can be regarded as curved in a way which is directly related to its mass and hence to the gravitational force itself. The singularity in spacetime may be regarded as where the curvature in spacetime becomes very high and you go beyond the classical theory of gravity, into the quantum regime. The event horizon surrounding the singularity functions as a one-way

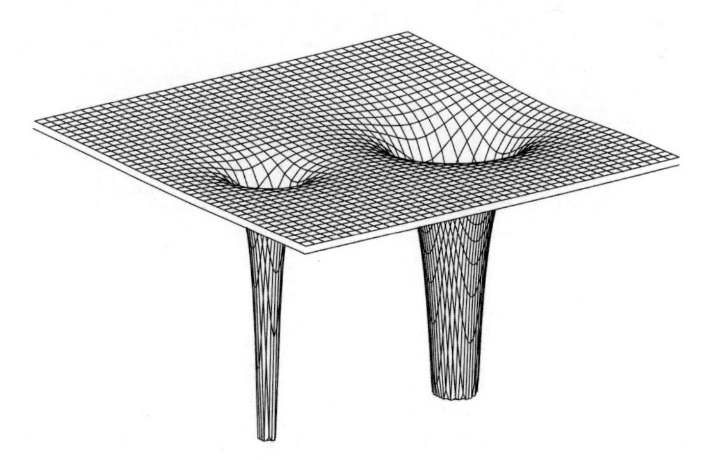

2. The distortion, i.e. curvature, in spacetime due to the presence of masses.

membrane: particles and photons can enter the black hole from outside but nothing can escape from within the horizon of the black hole out to the external Universe. In fact mass is not the only property that a black hole may possess and be measured by. If the black hole is rotating, that is to say it possesses some spin, then even more extreme behaviour emerges. Before we examine this, we will take a little detour to learn a little more about how we may schematically represent spacetime itself.

Chapter 2
Navigating through spacetime

Mathematics is the exquisitely perfect language needed for describing how the theory of relativity applies to the physical Universe and all of spacetime, and that description includes the strange behaviour that occurs near black holes. A mathematical description, while powerful and exact, even so can be something of a foreign and forbidding language for those without the appropriate technical training. Descriptive words, however eloquent, lack the rigour and power of a mathematical equation and can be imprecise and limiting. Pictures however, being (it is said) worth a thousand words, can be not only a useful compromise but a very helpful way to visualize what is going on. For this reason, it is well worth spending a little effort to understand a particular type of picture, called a spacetime diagram. This will help in understanding the nature of spacetime around black holes.

Spacetime diagrams

The cartoon in Figure 3 shows a simple spacetime diagram. Following tradition, the 'time-like' axis is the one that is vertical on the page and the 'space-like' axis is drawn perpendicular to this. Of course, we really need four axes to describe spacetime because there are three space-like axes (usually denoted x, y, and z) and one time-like axis. However, two axes will suffice for our purpose

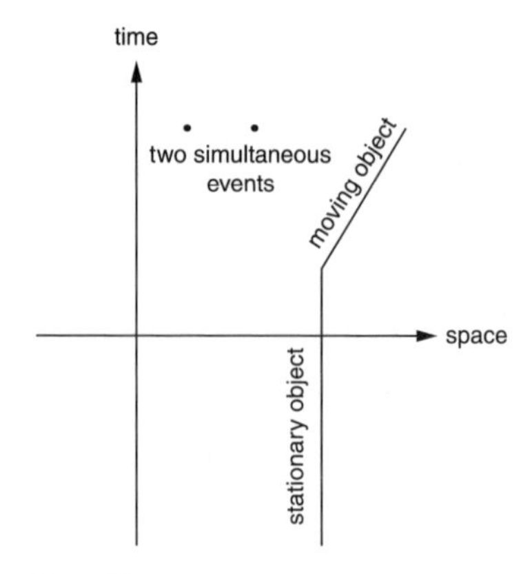

3. A simple spacetime diagram.

(and of course four mutually perpendicular axes are impossible to draw!). Where these two axes intersect is called the origin, and this may be regarded as the point of 'here and now' for the observer who has constructed their spacetime diagram. An idealized instantaneous event, say the click of a camera shutter, occurs at a particular moment in time and at a particular location in space. Such an instantaneous event is represented by a dot on a spacetime diagram, appropriate to the time and spatial location in question. There are two dots in Figure 3, which are spatially separated (they do not occur at the same point on the space axis) but they are simultaneous (they have the identical coordinate on the time axis). You could imagine these two dots correspond to the simultaneous shutter presses of two photographers who are standing some distance apart from one another, photographing the same spectacle. If points represent events, what do lines in a spacetime diagram represent? A line simply shows a path of an object through spacetime. As we live our lives, we journey through spacetime and the path we leave behind us (somewhat as a snail leaves a glistening trail of slime behind it) is a line in spacetime, and in the jargon this is called a *worldline*. If you stay at home all day, your worldline is a vertical path through spacetime (with space coordinate = '22 Acacia Avenue', for example). You move forward in time but are fixed in space. If on the other hand you made a long journey, your worldline slants over because your distance changes with time, because you move in space as well as time.

For example, look at the worldline shown in Figure 3, the line which is part vertical, then further up becomes slanting. This corresponds to the worldline of some other entity, which is stationary for the time indicated by the vertical extent of the line. An example might be a camera belonging to one of the photographers, left on a chair (so that its worldline is vertical because its position isn't changing), before it was stolen and whisked away (when the spatial location changes continuously). Where this line becomes slanting is where its spatial location is changing with time. The slope of this line tells you about the rate of change of distance with time, which is more commonly called the speed. In this case this is the speed at which the thief is whisking away the stolen camera. The faster the thief is making off with the camera, in other words the more ground he is covering in a given time, the less vertical and the more slanting this part of the line will be. There is of course a robust upper limit to the speed at which the thief can run off with his illegally gotten gains and this, as discussed in Chapter 1, is the speed of light. The trajectory of a beam of light would be represented by a maximally slanting line (commonly represented in spacetime diagrams as being at 45 degrees to the time axis by using cleverly crafted units). Because nothing can go faster than that speed, no worldline can be at a greater angle to the time axis than this.

Worldlines on a spacetime diagram having this maximally slanting angle, corresponding to this maximal speed, the speed of light, give rise to an important concept called a light cone. The idea of this is very simple: you can only have an effect on the Universe in the future by some prior cause and that causal sequence cannot propagate faster than the speed of light. Therefore your 'sphere of influence' right now is contained in a restricted range of

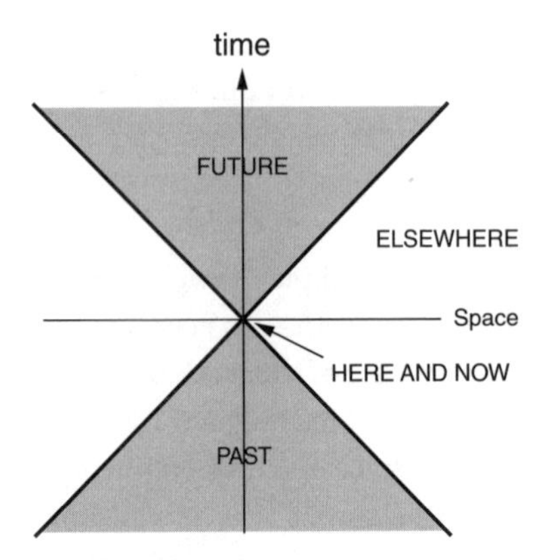

4. A simple light cone diagram.

spacetime, namely that part which is within a 45-degree angle to the positive time axis as shown in Figure 4. Moreover, you can only have been influenced by a causal chain of events that could not have propagated faster than the speed of light. Therefore only events within a 45-degree angle to the backwards time axis can influence you now. If we now draw a spacetime diagram with two space-like axes and one time-like axis, then the triangles in Figure 4 become cones and these are what we mean by *light cones*, as shown in Figure 5. The light cone in Figure 5 delineates regions of space within which an observer (deemed to be located at the origin, their 'here and now') could in principle reach (or have reached in the past) without having to invoke breaking the cosmic speed limit and travelling faster than the speed of light. The region centred on the positive (future) time axis is known as the future light cone while the cone centred on the negative time axis (i.e. past times) is known as the past light cone.

Thus the assassination of Julius Caesar in 44 BC is part of your past, because there is a conceivable causal link between that event and you. (If you had to learn about it at school, that demonstrates the existence of a causal link!) Because light from the Andromeda Galaxy can reach a telescope on Earth, it too is part of your past. However, the light takes 6 million years to get to us, so it is the

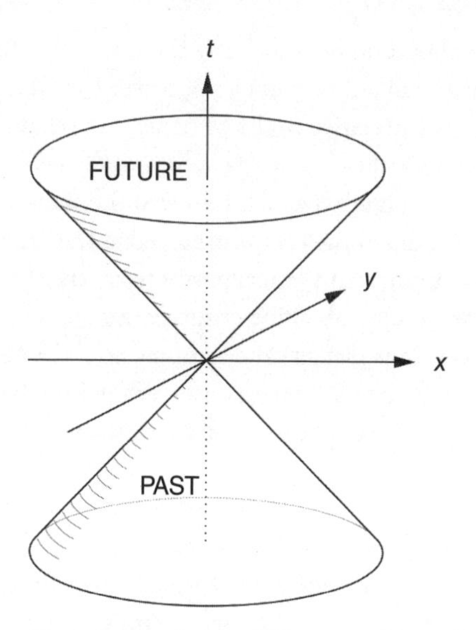

5. A spacetime diagram showing the light cone of a particular observer.

Andromeda Galaxy of 6 million years ago that is part of your past and sits on your light cone. The Andromeda Galaxy of today, or even the Andromeda Galaxy of 44 BC, is outside your light cone. Events happening on Andromeda, either now or even back in 44 BC, cannot influence you right now because any causal link would have had to travel faster than the speed of light.

The three spacetime diagrams that we have seen in this chapter so far have their axes labelled as time and space. In fact, professionals wouldn't normally include axis labels or even the axes in spacetime diagrams. This isn't simply that it is so routine that time goes up and space goes across that professional astrophysicists get sloppy (though that's not an unknown phenomenon) but it is because the exact positions in spacetime cannot be agreed upon by all observers. In the world of special relativity, the notion of simultaneity breaks down. Just because two events are seen to be simultaneous for one observer doesn't at all mean that they are simultaneous for other observers.

Thus the two photographers pressing the shutters of their cameras 'simultaneously' will not be what an observer travelling in a spacecraft very fast relative to the cameras sees. That observer will deduce one camera click occurring substantially before the other. The two points in Figure 3 which I drew at the same vertical height (since I claimed the events occurred at the same time) would appear at different vertical positions on the spacetime diagram of the rapidly travelling observer. Einstein's relativity insists her diagram is just as valid as mine. So if the points on a spacetime diagram depend on an observer's point of view, i.e. their frame of reference, what's the reason for drawing them?

To understand this, it is helpful to focus on the worldline of a moving particle and so we will now draw a new spacetime diagram in which a particle moves through spacetime, taking its light cone with it (this trick is known as working within the co-moving frame). Notice that in Figure 6 the particle's path (i.e. its worldline) always stays within the light cone as it cannot travel faster than the speed of light.

Einstein's Special Theory of Relativity, which is a subset of his General Theory, pertains to a restricted set of physical situations. A different conceptual framework beyond Special Relativity is needed in the context of spacetime which is expanding, the pre-eminent example of which is the expanding Universe. In this context, the manifestation of causality is such that you cannot move faster than the speed of light with respect to your local bit of space.

How do objects know where to go?

Although photons have no mass, it turns out that they are still influenced by gravity. It is best not to think of this as due to a force, but rather that this comes about because of the curvature of spacetime. A photon is usually thought to travel in a straight line, which is where we get the notion of a 'light ray'. However, through a curved spacetime it will follow a path known as a *geodesic*.

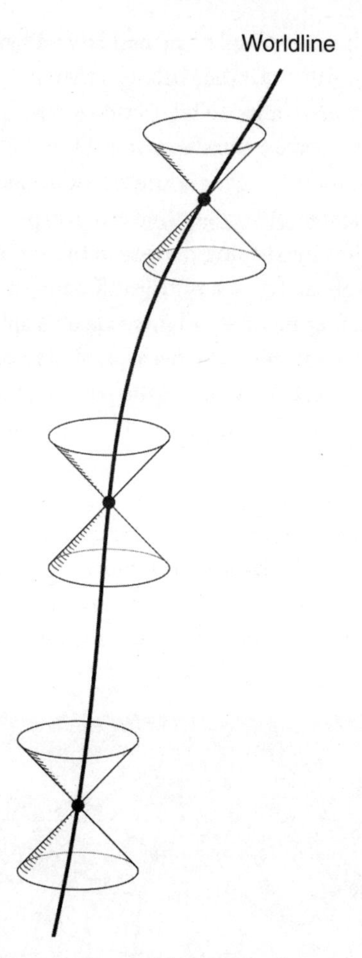

6. A spacetime diagram of a particle moving along its worldline, that is always contained within its future light cone.

Despite its Earth-based connotations, a geodesic (whose name comes from geodesy, i.e. measuring the lie of the land of our planet's surface) is an important concept describing the nature of spacetime throughout the Universe. If space were not curved (meaning entirely consistent with everyday geometry that we may have learned at school from Euclid or one of his successors), then a geodesic would be the 'straight line path' that a light ray would travel. But the shortest distance between two points, which is the route that a light ray 'wants' to take, is known by the term 'null geodesic'. In curved space the shortest distance between two points isn't what we think of as straight, but 'geodesics are straight lines in curved spaces'. A straight line can also be characterized as the path you follow by keeping moving in the same direction. An example of how geometry is seriously different on a curved surface comes from considering lines of longitude on a sphere. Two adjacent lines of longitude (which are parallel to one another at the equator) will meet at a point at the pole, as shown in Figure 7. However, in flat space parallel lines will meet only at infinity (as per Euclid's last axiom).

Actually, where spacetime is curved, for example because of the presence of mass, that curvature is manifested in the path that a light ray or (a mental device used by physicists) a 'test particle' freely able to move with no influence of any external force, would

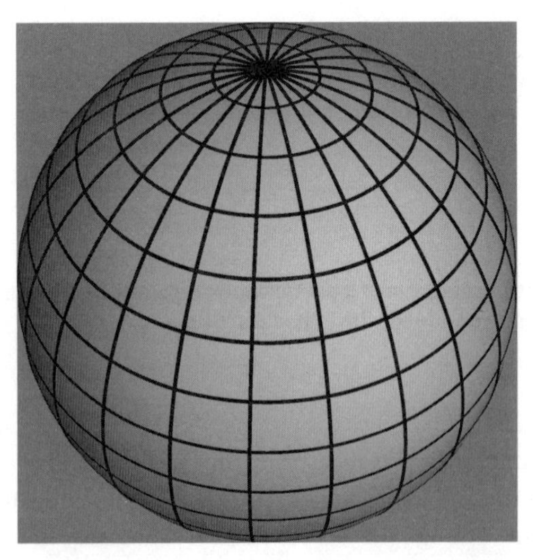

7. Lines of longitude on a sphere are parallel at the equator, and meet at a point at the poles.

move along between two events. Two events should be regarded as two points in 4-D space time, each denoted in the form (t, x, y, z).

A rule called a *metric* tells us how clocks and rulers measure the separations between events in space and time and provide the basis for working out problems in geometry. A very simple example of a metric is Pythagoras' theorem, which tells us how to compute the distance between two points that lie in a plane. The solutions to Einstein's field equations tell us how to calculate the metric of spacetime when the distribution of matter is known. We use this to construct the geodesics for the real Universe. For example, one of the first pieces of observational evidence for General Relativity was the bending of starlight by the Sun, measured during a solar eclipse (a good time to examine the apparent positions of stars close to the Sun's disc because light from the disc is blocked out by the Moon, an opportunity seized upon by Sir Arthur Eddington in 1919). The Sun's mass curves spacetime. Thus the shortest path (the geodesic) from a distant star to a telescope on Earth is not quite a straight line: it is bent round by the Sun's gravitational field, as shown in Figure 8.

The bending of starlight demonstrates that space is curved, but Einstein's General Theory tells us it is actually spacetime that is curved. Therefore we might expect that mass also has some

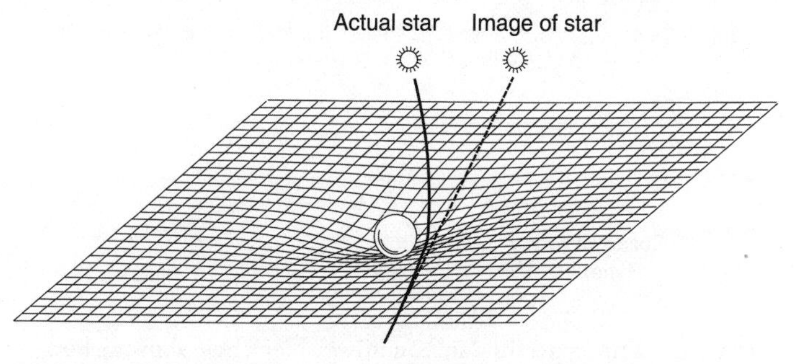

8. A mass such as the Sun causes distortion, or curvature, in spacetime.

strange effects on time. In fact, even the Earth's gravitational field is sufficient to make Earth-bound clocks tick a bit slower than they would do in deep space, although the effect is small (roughly one part in a billion) but measurable. The gravitational effects near the event horizon of a black hole are much stronger. Thus, even for the simplest case of a non-spinning black hole, time runs differently close to the black hole compared to how it runs at a huge distance from the black hole. This is a real effect and does not depend on how the time is measured (for example by an atomic clock, or by a digital watch). It follows directly from the curvature of spacetime induced by the mass which tips the light cones towards the mass. Figure 9 indicates the general effect.

Black holes profoundly affect the orientations of the light cones. As a particle approaches a black hole, its future light cone tilts

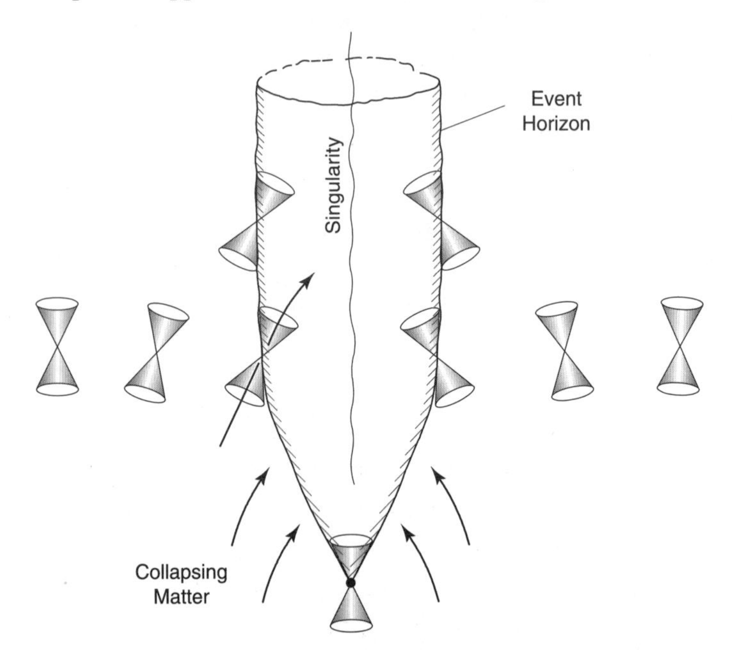

9. Diagram of the spacetime surrounding a black hole showing how the future light cones for objects on the event horizon lie inside the event horizon.

more and more towards the black hole, so that the black hole becomes more and more a part of its inevitable future. When the particle crosses the event horizon, all of its possible future trajectories end inside the black hole. Just within the event horizon, the light cone tilting is so great that one side becomes parallel with the event horizon and the future lies entirely within the event horizon; escape from the black hole is not possible. Figure 9 also illustrates this point: it is essentially a representation of 'local spacetime diagrams', because the assembly of light cones allows you to understand the local conditions experienced by a test particle located at different positions. In this figure, time increases up the page and so this diagram also gives a sense of how a black hole forms and grows due to infalling matter.

Just as for the dark stars of Michell and Laplace discussed in Chapter 1 which could have sustained planetary systems in orbit around them much like our Solar System, so it is that we only know that a black hole is nearby due to its gravitational pull. This might lead you to think that the only property that characterizes a black hole is its mass. In fact, whether or not a black hole is rotating has a dramatic effect on its properties, and I will explain how this comes about in Chapter 3.

Chapter 3
Characterizing black holes

In Chapter 1, we introduced the concept of a mass singularity, forming in gravitational collapse, and surrounded by an event horizon. Examples of such objects that are not spinning are called *Schwarzschild* black holes and this term specifically denotes black holes that are not rotating: in the jargon, they have no *spin*. Simply put, the only characteristic that distinguishes one Schwarzschild black hole from another (other than location) is how massive it is. In Chapter 7 we will learn how black holes grow but for now, it will suffice to know that collapse under gravity is the key ingredient. If there is any rotation whatsoever in the pre-collapsed matter, however gentle, then as the collapse occurs the rotation rate will increase (unless something acts to stop that happening). This arises due to a remarkable physical law known as the *conservation of angular momentum*. This law is illustrated by a pirouetting skater: as she pulls her arms in she spins faster. In the same way, if the star that gives rise to the black hole is gently rotating then the black hole that it ultimately forms will be spinning significantly and is termed a *Kerr black hole*. Most stars are in fact rotating, because they themselves are formed from the gravitational collapse of slowly rotating massive gas clouds. (If such a gas cloud had even a minute amount of net rotation then the collapsing cloud will have non-zero angular momentum, and as the matter occupies an increasingly smaller volume the final rotation of the collapsed object may well be rather rapid.) Thus we see that rotation, more commonly called spin, is likely to be a prevalent, if not actually a ubiquitous, characteristic for black holes that have just formed from the collapse of matter. We now believe that spin is as inevitable in real astrophysical black holes as it is in current-day politics (though in the latter case it arises from something other than the conservation of angular momentum!).

We have now stated that a second physical parameter, that of spin or angular momentum, is a characteristic that distinguishes one black hole from another just as mass does. Thus, there are two properties of black holes that are important to keep in mind as we study the behaviour of black holes: mass and spin. In principle, there is a third characteristic of black holes that might be relevant to their behaviour: electrical charge. This is also a conserved quantity in physics, and the forces between electric charges, known as electrostatic forces, have a number of resemblances to gravitational force. A key similarity is that both are (on large scales) examples of inverse-square laws meaning that, in the case of two massive objects, as you double the distance that separates them from one another the gravitational force they experience reduces to a quarter of the original value. A key difference is that while gravity is always attractive, electrostatic charges are only sometimes attractive (when the two bodies are oppositely charged, i.e. one is positive and the other is negative). They are at other times repulsive (when the bodies have charge of the same sign, either both positive or both negative, they repel each other). If two charged bodies have the same type of charge, then electrostatic repulsion will tend to prevent them coalescing, even if gravity is tending to attract them. So while charge could in principle be a third property of black holes that one might hope to measure, in reality a charged black hole would be rapidly neutralized by the surrounding matter. It is therefore a good operational assumption that there are only two relevant properties of black holes that distinguish one from another: mass and spin.
That's all!

Now, you might wonder whether black holes could be distinguished by their composition. One might have been formed from a hydrogen gas cloud, another from a helium gas cloud. Why should it be that the provenance of the collapsed matter that gave rise to the black hole isn't manifested in the measurable properties of the black hole subsequently formed? That's because information can't get out of the event horizon! Light is the means by which information might be transmitted, but we have already seen in Chapter 1 that it cannot escape from inside the event horizon of a black hole. Thus the chemical composition of the matter that fell into the black hole can have no effect on the properties of the black hole as determined from the outside. It would not be correct to think of gravity as something that needs to 'get out of' the black hole. The continued existence of a gravitational field external to the black hole is something that is laid down in the formation of the black hole as spacetime becomes distorted. No influence from inside the black hole could change the external field after the event horizon has formed.

Black holes have no hair

When we are asked to describe another person, a distinguishing characteristic that is often included is their hair (for example, strawberry blonde, or silver grey or chocolate brown). There are sometimes clues in the nature of people's hair as to their age or their nationality. Information about further physical characteristics such as 'Body Mass Index' might provide information on their diet. In contrast to humans, black holes are entities that have absolutely no distinguishing characteristics other than their mass and their spin (neglecting charge for the reasons noted above). This is captured in the breviloquent phrase 'Black holes have no hair', coined by John Wheeler to emphasize that there is nothing about a black hole that bears any evidence of the nature of its progenitor star. Not its shape, not its lumpiness, not its landscape, not its magnetism, not its chemical composition. Nothing. Calculations done by, amongst others, the Belarusian physicist Yakov Zel'dovich demonstrated that if a non-spherical star with a lumpy surface collapsed to form a black hole, its event horizon would ultimately settle down to a smooth equilibrium shape having no lumps or bumps of any kind. So, a black hole never has a bad hair day! The only things you can know about it are its mass and spin.

Spin changes reality

Perhaps the most remarkable feature of a spinning black hole is that the gravitational field pulls objects around the black hole's axis of rotation, not merely in towards its centre. This effect is called *frame dragging*. A particle dropped radially onto a Kerr black hole will acquire non-radial (i.e. rotating) components of motion as it falls freely in the black hole's gravitational field.

What this means for a test particle having spin (such as a small gyroscope) is that if it falls freely towards a rotating massive body, such as a Kerr black hole, it will acquire a change to its spin axis. It is as though its local frame of reference was dragged by the rotation of the central massive body. This phenomenon, discovered in 1918, called the Lense–Thirring effect actually occurs not just around black holes, but to some extent around any spinning object. If you put a very precise gyroscope in orbit around the Earth, the frame dragging causes the gyroscope to precess.

It is Einstein's field equations that describe the mathematics of black holes and, as also mentioned in Chapter 1, Karl Schwarzschild solved these equations for the case of the stationary (non-rotating) black hole, a remarkable achievement given that he did this in 1915, the same year that Einstein introduced his general theory of relativity. The case of the spinning black hole was treated much later by New Zealander Roy Kerr in 1965. A few years after this, the Australian Brandon Carter explored Kerr's solution further still. Carter carried out an in-depth investigation into the consequences of the Kerr metric. He established that a spinning black hole causes a dramatic swirling vortex in the spacetime that surrounds it which arises because of the dragging of the reference frame. An example of a vortex is a whirlwind—close to the centre of the whirlwind the air swirls rapidly, carrying with it anything in its path, be it hay in a hay field or sand in a desert. Further from the whirlwind the air (and hence hay or sand) rotates much more slowly. So it is too, with spacetime surrounding a spinning black hole: far away from the event horizon, the speed at which spacetime itself rotates is slow, but at the horizon, spacetime itself spins with the same speed that the horizon spins.

The event horizon for the spinning (Kerr) black hole is much the same as for a non-spinning (Schwarzschild) black hole, except that the faster the black hole is spinning, the deeper the gravitational potential well: a Kerr black hole forms a deeper gravitational potential well than a Schwarzschild black hole of the same mass, and therefore a Kerr black hole can be a more powerful energy source than a non-spinning one, a point to which we return in Chapter 7. In the meantime, it is helpful to summarize this behaviour by saying the event horizon of a Schwarzschild black hole depends only on mass, but that of a Kerr black hole depends on both mass and spin.

An outstanding question is whether there could be, even in principle, any spacetime singularities that are not enclosed within and hidden by event horizons—a so-called 'naked singularity'. By definition, all black hole solutions to the Einstein field equations do have event horizons and, as shown in Chapter 1, no light and therefore no information can escape from within such horizons. All black hole singularities are believed to be enclosed within event horizons and therefore not 'naked', so that direct information about the singularity is inaccessible from the rest of the Universe. The so-called cosmic censorship conjecture was formulated by the British mathematician Roger Penrose and states that all spacetime singularities formed from regular initial conditions are hidden by event horizons and that there are no naked singularities out in space.

How much spin is too much?

There is a limit to how much angular momentum a black hole can have. This limit depends on the mass of the black hole, so that a more massive black hole can spin faster than a less massive black hole. A black hole that is rotating close to this maximum limit is known as an extreme Kerr black hole. It is possible to show that if you try to spin up a black hole, to make an extreme Kerr black hole, by firing rapidly rotating matter into it (i.e. giving it a stir) then centrifugal forces prevent the matter from even entering the event horizon.

Somewhat further out from the event horizon of a rotating black hole is another significant mathematical surface which is known as the *static limit*. The dragging of inertial frames means that if the spin of the massive body is non-zero then there are no stationary observers inside of this surface: every physically realizable reference frame inside the static limit must rotate. Within this surface, space is spinning so fast that light itself has to rotate with the black hole, i.e. it is impossible to remain motionless. The region between the static limit and the event horizon is known as the *ergosphere*, which rather confusingly is not spherical, as shown in Figure 10. In equatorial directions the ergosphere is much larger than the event horizon, but in the polar directions the radius of the ergosphere is the same as the radius of the event horizon. The resulting shape of the ergosphere is an oblate spheroid, resembling the shape of a Jarrahdale pumpkin (without the stalk). The first two syllables of ergosphere, however, come from the Greek noun *érgon* relating to 'work' or 'energy' (as in 'ergonomics') from which the old unit of energy, the erg, is also derived. It is intriguing to note that in addition there is a Greek verb *ergo* which means to enclose and keep away, appropriately for the nature of the ergosphere. Perhaps this may have been in the

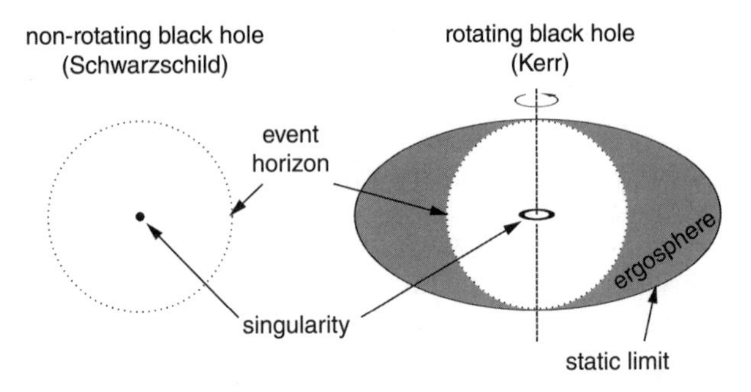

10. The different surfaces around a Schwarzschild (stationary) black hole and around a Kerr (spinning) black hole (in the frequently used representation of 'Boyer-Lindquist' coordinates).

minds of Roger Penrose and Demetrios Christodoulou who coined and championed the name of this region around a spinning black hole. The importance of the ergosphere is that it is the region within which energy can be extracted away from the black hole.

Since inside the ergosphere space is spinning, particles of matter within that space also get swept up into a rotational motion. Considerable rotational energy is therefore stored in this rotation of space, a very important point to which we return in Chapter 8.

White holes and wormholes

Einstein's equations of General Relativity are particularly rich and allow many different solutions describing alternative versions of curved spacetime. This provides an almost inexhaustible source of possible universes for cosmologists to describe and think about. Which type of universe we actually live in is a matter that can only be decided by observation (if at all!). But that doesn't stop mathematical physicists playing around with Einstein's equations to find all kinds of interesting solutions.

One intriguing object that can be dreamt up by mathematical physicists is what is called a *white hole*. A white hole behaves just like a black hole but with the direction of time reversed (imagine a movie played backwards). Instead of matter being sucked in, it is spewed out. Instead of the event horizon marking out the region from which you can never escape, it stakes out the region into which nothing could ever enter. Once matter exits from a white hole, it can never return there; its entire future is outside. As we see in Chapter 6, a black hole is formed from a collapsing star and must eventually evaporate by the laws of quantum mechanics into Hawking radiation (see Chapter 5). A white hole, on the other hand, could only result from radiation that for some reason spontaneously assembles into a black hole. It is not easy to understand how this could happen in practice, and moreover Douglas Eardley has demonstrated that white holes are inherently unstable.

When Einstein and his student Nathan Rosen were playing around with Einstein's equations in the 1930s, they found an interesting solution. If a region of spacetime could be strongly curved, it might be possible for it to become sufficiently folded that two parts of spacetime which had previously been separated by a large distance could become connected by a small bridge, or wormhole, as shown in Figure 11. The enormous distances between the stars and galaxies have always been unfavourable for those writers who wish to set human dramas on a cosmic stage, and wormholes (also known as Einstein–Rosen bridges) have provided the perfect plotting device for writers to transport their heroes and villains about. This mathematical invention has been an absolute boon to the writers of science fiction, because it provides a ready means for traversing enormous distances through space and thereby to sustain various highly artificial and unbelievable plot devices. Yet again, we have no observational evidence that wormholes actually exist in our Universe. In addition, there is considerable theoretical evidence that a wormhole, once formed, would not be stable for very long. It seems that to keep a wormhole propped open, one needs a large amount of negative energy matter, and all normal matter has

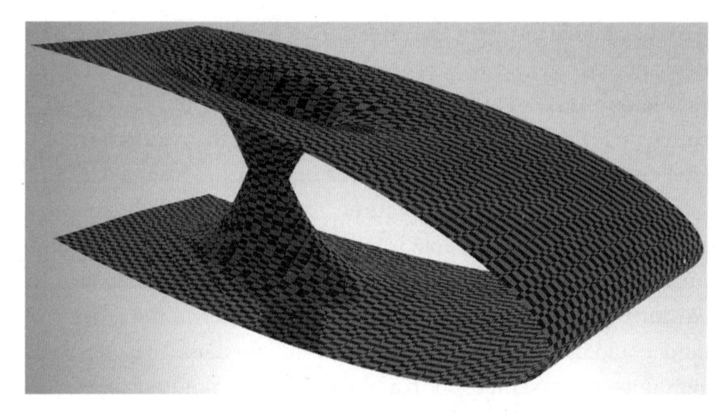

11. A wormhole connecting two otherwise separate regions of spacetime.

positive energy (this is connected with the fact that gravity is normally always attractive). Normal matter passing through a wormhole may be enough to destabilize and destroy it, causing it to turn into a black hole singularity.

If wormholes did exist, and could be maintained for any reasonable length of time, they would have some surprising and bizarre properties. Not only would they provide a means for taking an enormous shortcut across a vast expanse of space, but they would also allow a traveller to journey back in time. One would then be able to construct closed time-like curves, loops in spacetime in which the light cones form a ring (see Figure 12) so that, like in the movie *Groundhog Day*, a person travelling along a closed time-like curve would simply repeat their same experiences over and over again.

In fact, there are a number of solutions to Einstein's equations in addition to wormholes which have this alarming and counterintuitive property. In 1949, the mathematician Kurt Gödel found a solution that described a spinning universe, and this contains exactly the same sort of closed time-like curves which pass through events again and again in an endless *Groundhog Day*

12. A closed time-like loop, on which your future becomes your past.

cycle. (Evidently 'free will' is not part of the field equations!)
The part of the Kerr solution thought to have genuine physical significance in the real world is that which describes the spacetime outside of the event horizon. However, it is unclear whether the part of the Kerr solution inside the event horizon, while mathematically sound, has any physical relevance. In this part of the Kerr solution, the singularity is not a point (as it is for the non-rotating black hole) but has the form of a rapidly rotating ring (however, the physical validity is very speculative). This ring-like singularity is surrounded by closed time-like curves. On such a curve, your future is also in your past and you have the theoretical possibility of murdering one of your own grandparents before they had produced your parents! Thus the existence of closed time-like curves seems to create the possibility of all kinds of paradoxes relating to time travel. One possible solution to this is to admit that we do not have a theory that links quantum mechanics (which describes the very small) and general relativity (which describes the very massive), in other words a theory of quantum gravity. We don't know the physics of extremely massive but very small objects. Most physicists think we need this to fully understand the behaviour of spacetime very close to singularities. Thus it may

be that these strange solutions to Einstein's equations do not actually occur in the Universe because they are prohibited by its fundamental quantum mechanical nature. Quantum effects may, for example, destabilize wormholes. Stephen Hawking believes this to be the case and has called this principle the 'Chronology Protection Conjecture'. He has quipped that this is the underlying principle that keeps the Universe safe for historians.

There is much about the interior of rotating black holes that pushes our understanding of fundamental physics to the limits and therefore to where much of our description is highly speculative. By contrast, the rotation of black holes and their effect on their surroundings is something that has enormous practical significance for understanding what we can see with our telescopes. Thus our next step is to consider in more detail what happens to matter when it falls into a black hole.

Chapter 4
Falling into a black hole . . .

How close is too close?

Before we can consider in detail what would happen if you or your belongings had the misfortune to fall into a black hole, it is important to understand the effect of an observer's particular perspective, or frame of reference. This means that different observers see very different things. Exactly what your perspective is on an object falling into a black hole depends on how far away you are from that object (and indeed whether you are that object!). Consider a particle of light, a photon, that is outside the event horizon of a black hole: since it is outside the horizon, it can in principle escape. Inside the event horizon it would be a different story—the photon could not escape the gravitational field of the black hole. But even outside the event horizon, a photon that is travelling away from the black hole will not escape completely unscathed. The photon suffers a loss in its energy due to the work it has to do against gravity. This is an example of a gravitational potential well; just as energy would be needed to haul yourself upwards out of a deep well, so the photon needs to expend energy to pull itself away from the region near a massive object. The effect has even been measured for photons moving in the Earth's gravity. The energy of a photon is inversely related to its wavelength: a high-energy photon has a short wavelength whereas a low-energy photon has a long wavelength. The photon loses energy as it retreats away from the black hole, so its wavelength increases. This changes the colour of the light, moving from the blue (short wavelength) towards the red (long wavelength) end of the spectrum (this effect is called *redshift*). This sort of redshift, known as *gravitational redshift*, arises where spacetime itself stretches out, or is curved, for example by the effect of a massive body such as a black hole. Note that John Michell, despite having significant original thoughts about dark stars, was incorrect in thinking that the velocity of light decreases as it climbs out of the potential well. We now know that it is the wavelength (hence frequency) of light that is affected by the presence of a massive star.

What happens to time near a black hole?

In Chapters 1 and 2 I described how spacetime is distorted by the presence of a mass (i.e. something which produces its own gravitational field) and this means that not just space, but also time is affected close to a black hole.

Imagine you want to keep a safe distance from a Schwarzschild black hole but you want to learn more about how time behaves nearby. Thus you have arranged for twenty-six fixed observers to be stationed close to the black hole's event horizon but definitely safely outside it. These observers are named A to Z, and are arranged in a line with A closest to the event horizon and with Z being nearest to you, safely far away. Each observer from A to Z has a good clock with which to measure their local time, at their particular location. As part of the deal to persuade A to Z to participate in this experiment, you had offered them each an inducement in the form of a gift of an additional, unusual clock that had been adjusted so that the time on it would read the same as the time on your clock at your safe distance. Participant Z, closest to you, would find that the two clocks in his possession read slightly different times because his own clock, which measures local time ('*proper time*' in the jargon), would be running slightly more slowly than the gift clock which matches the time you measure at your rather safer and more remote distance. The collated results of participants Z to A would display a remarkable effect: closer to a black hole, a clock measuring time 'runs more slowly' compared with the distant time as reported on the participants' specially adjusted gift clocks. This effect, described by Einstein's theory of general relativity, is known as *time dilation*. The effect would be greater and greater for the observers nearer the start of the alphabet who are nearer to the black hole. The greater the proximity to a black hole, the more slowly a local clock (of any kind: atomic, biochemical) will run compared to a clock used by a distant observer.

Suppose you were multi-tasking your experiments with a different set of twenty-six observers at the same distances from a different black hole. They are arranged in just the same way as their namesakes near the first black hole. However, in this second case, the black hole has twice the mass of the black hole in your first experiment. The unusual clocks you had prepared as gifts for this second set of observers would need to be radically altered as for your original experiment, but the rate at which each unusual clock has to be adjusted is exactly double that of the rate needed for the corresponding clock in the first set of gift clocks at the exact same distance from the centre of the first black hole which has half the mass of the second. These time dilation effects are larger if the black hole mass is larger, and also become more extreme the closer you get to the event horizon.

Note that this time dilation is not a consequence of some additional light-travel time for a clock closer to the black hole and hence further from you, the safely-distant observer: there is not merely a time offset for an observer further away from the black hole. The closer a clock is to a black hole, the slower is the rate at which time is measured to flow on that clock, no matter what reputable means you use to measure that flow of time. Time itself is stretched (or, indeed, dilated).

What is the corollary of time dilation near a black hole? This causes effects that happen in the frame of an observer very local to the black hole to be measured to be very different from those in the frame of an observer who is very distant, worlds apart in fact.

Let's now consider what happens if in your first experiment, observer A became a little careless and dropped his first clock (the one with which he could measure proper time at his location) so that it fell towards the black hole. Despite this disaster, he would be nonetheless safely gripping onto the gift clock with which you had enticed him to participate in the experiment. Both you and A would see his first clock move towards the hole. The clock would find itself moving into the black hole, more and more rapidly. You and A would gradually notice that the time you read on the plummeting clock becomes even more discrepant with the time on A's other clock (namely the clock that was adjusted to run faster than the local clock in order that it would read the same time as the one corresponding to your time). After a while both you and A would begin to notice that time stops for the plummeting clock. A photon emitted at the event horizon towards a distant observer appears to stay there indefinitely. What happens to anything that falls into a black hole after it has passed within the critical radius of the event horizon is unknowable to an external observer. So the event horizon may be regarded as a hole in spacetime. No light will emerge from within the event horizon, as we saw in Chapter 1. That is why it is black. However, in the reference frame of the dropped clock plummeting through the event horizon, life is very far from unchanging. From the clock's perspective, it would travel to the singularity in a mere one ten-thousandth of a second, assuming that the black hole had a mass of ten times that of our Sun. If the clock had the misfortune to fall into a supermassive black hole with a mass one billion times that of our Sun (such as we meet when we study quasars in Chapter 8), its journey time inwards between the vastly larger event horizon and the singularity would be a more leisurely few hours.

Tidal forces near a black hole

Suppose in a weak moment, person A wonders about jumping, feet first towards the black hole, in hopes of being reunited with the clock he dropped. What would happen? Such a leap would prove to be a big mistake, as the survival outcome would be zero. The difference between the gravitational force on his feet and the force on his head would become extreme. This is a feature of any inverse-square force field, such as gravity from a massive body. The Earth is rather a long way from the moon, yet even the small differences in gravitational force due to the moon experienced on opposite sides of the Earth, known as tidal forces, are at the root of why the tides come and go about twice per day. In general, these forces resulting from differences in gravity in different places are called tidal forces. There are additional factors that enrich the details of the rising and falling of tides such as the gravitational force due to the relative angle of the moon, and the detailed shapes of continental masses. But even if the surface of Earth were entirely covered by ocean without land, there would still be tides with the amplitude of the sea level varying by about 20 cm twice per day, simply because of the differential gravitational force experienced by points on the planet at different distances from the Sun.

Let's now consider the smaller distance between me and the centre of the Earth. As I sit typing this chapter, my head is somewhat over a metre higher than my feet which are on the floor of my study. My feet are thus closer to the centre of the Earth than my head is. Because the gravitational force follows an inverse-square law behaviour as though all the mass of the Earth were located at the very centre of the Earth, and because my feet have a smaller distance to this centre they feel a stronger force, or pull, to the centre of the Earth than my head does. But actually, the difference is rather slender: for a height difference of one metre the difference in gravitational force is three parts in ten million. This is such a slight difference because I am about

6,400 km from the centre of the Earth. Much closer to a point mass such as a black hole, the difference in gravitational force experienced at points just a metre apart in the direction towards the black hole would be vastly more extreme. So extreme that close to the singularity A's feet would be stretched away from his knees and the rest of his body beyond what his tendons and muscles could hold together, and he would be elongated into something resembling long spaghetti. Best not to jump.

Dynamic spacetime

The rotation of a black hole makes an important difference regarding how close matter can orbit around it, and this relates to how much energy can be extracted from it. From the work of Roy Kerr and his solution to the Einstein field equations, we know that the smallest orbit that a particle can have around a black hole without falling in depends on just how fast the hole is spinning. The faster a black hole is spinning, the closer the matter can get before the hole swallows it, as illustrated in Figure 13. If you drop something straight down into a spinning black hole, it will start orbiting the hole even though there is nothing but empty spacetime outside the hole. Outside the ergosphere, it is possible

13. Gas can orbit closer to a spinning black hole than to a non-rotating one.

to overcome this frame dragging using rockets, but not inside it. In the region inside the rotating black hole's ergosphere, just outside its event horizon, nothing can stand still. The spinning hole actually drags the spacetime and hence its contents around with it. A further aspect of this frame-dragging is that even if light itself is going against the direction the black hole is rotating, it will be carried in the reverse direction around the hole.

Orbiting around a black hole

It is interesting to ponder what would be the sequence of events if our Sun were to spontaneously metamorphose into a black hole right now. The first that you or I could know about it would be eight minutes later; the beautiful Spring sunlight by which I am writing would come to an abrupt halt. Although the luminosity of the single star we call our Sun is tiny by comparison with the quasars and microquasars discussed in Chapter 8, it is sufficiently close to the Earth that it provides on average about a kilowatt per square metre of power to our planet. Remarkably, this has been enough to sustain all life on the planet, allowing plants to grow and then be eaten by animals that are then eaten by other animals. The Sun has been the engine behind it all. But if fusion ceased in the Sun and it were (contrary to all expectation) to collapse into a black hole, then it would go very dark and we would all eventually die. (This is a bit of a gloomy outlook, but I encourage the reader to hold fast until Chapter 7, where we learn that our Sun is not the kind of star to form a black hole—it's too lightweight for that.) However, dynamically speaking, as far as planet Earth and the whole Solar System of planets, dwarf planets, and asteroids are concerned, nothing will change at all. All massive bodies in orbit around the Sun will continue in pretty much the same orbits. The way that gravity works is that whether the Sun has the same extent that it has now, or whether it collapses to a singularity within an event horizon of 3 km, the gravitational attraction outside the Sun would remain unchanged. The spherical collapse under gravity to a black hole would not change the angular momentum of the orbiting bodies at all, so the patterns and progressions and tides within the Solar System would continue utterly unaltered by the lack of sunshine.

Some new orbits would be possible however, much closer to the black-hole Sun than were possible previously when the solar plasma was in the way. However, these orbits could not get too close to the event horizon. The details of the warping of spacetime by a mass singularity mean that it is not possible to orbit just outside the event horizon itself. Attempting a circular orbit there would require corrective action by rockets in order to maintain the orbit. In fact, the mathematics shows that the closest that we or any other mass particle could exist on a stable circular orbit near a stationary black hole would be at a distance three times that of the Schwarzschild radius away. You have been warned.

Actually, unstable circular orbits are possible up to half this distance away from a Schwarzschild (non-spinning) black hole. This distance defines a spherical surface that is sometimes called the *photon sphere*. Even for a photon, these orbits are unstable, and before too long an orbiting photon would either slither in towards the black hole, never to return, or indeed away into space.

For a Kerr black hole though, one that has spin, the situation is different for the orbits near the black hole. In particular, there are two photon spheres, in contrast with the one photon sphere around a stationary Schwarzschild black hole. The outermost sphere is for photons that are orbiting oppositely to the direction of rotation of the black hole (the ones we say are on *retrograde* orbits). Inside this is the photon sphere for photons travelling in the same sense around the black hole as it is rotating (on *prograde* orbits). For a very slowly rotating black hole that isn't so very different from a Schwarzschild black hole, these two photon spheres are very nearly co-spatial. For black holes of increasing spin, these surfaces are increasingly further apart.

Moving closer in towards a rotating black hole, there is another important surface (discussed in Chapter 3), called the *static limit*. This is the surface at which nothing can remain static with respect to a distant observer: it is just impossible to sit still this close to a rotating black hole, no matter how powerful the rockets you might be equipped with. At this surface, even retrograde light rays are dragged along in the direction of rotation. It is still possible to escape from this close to a rotating black hole, with sufficient propulsion, but it's just not possible for anything to remain stationary and non-rotating here. Moving inwards still further, the next surface of significance is the event horizon we met in Chapter 1, the one-way membrane that we met originally in the context of Schwarzschild black holes. Crossing this outwards isn't possible and crossing it inwards has an ineluctable destiny, just as for the static black hole.

An orbit around a Kerr black hole is not generally confined to a plane. The only orbits confined to a plane are those in the plane that contains the equator (i.e. the plane of mirror symmetry of the spinning black hole). Orbits out of this equatorial plane move in three dimensions. These orbits are confined to a volume that is limited by a maximum and minimum radius and by a maximum angle away from the equatorial plane.

The details of the spin of a black hole have a dramatic effect on how close particles may encounter the black hole, which itself depends on their direction of travel relative to the spin. For a maximally spinning black hole, the photon sphere for light rays orbiting in the same sense (prograde) as the black hole spin has a radius that is half of what the Schwarzschild radius would be. For light rays on retrograde orbits, the radius of their photon sphere is twice the Schwarzschild radius. For particles with mass that are on prograde orbits, the innermost stable circular orbit on which they can move is again at half of the Schwarzschild radius. For those on retrograde orbits, such a close distance would be unstable: their innermost stable circular orbit is at 4.5 times the Schwarzschild radius. Thus, a rotating black hole enables particles on prograde orbits to orbit more closely without reaching the point of no return at the event horizon, more closely than if the black hole were non-rotating. In Chapter 7, we consider the importance of just how close matter can orbit before falling onto a black hole and how much energy may be consequently leveraged.

Chapter 5
Entropy and thermodynamics of black holes

You are what you eat

It is often said that you are what you eat. Thus if your diet is purely junk food and chocolate, then your complexion, not to mention your physical and mental well-being, will be rather different than if you subsist on a healthy diet of salad and Mediterranean food. However, it seems that black holes are not fussy eaters. Whether they are hoovering up a vast expanse of interstellar dust or a cubic light-year of fried eggs, their mass will similarly increase inexorably. In fact, after a black hole has finished its sumptuous meal, you have no way of telling what it was eating, only how much it has consumed (although you could tell if what it ate had charge or angular momentum). You only know the quantity of its diet, not about the quality. The 'no-hair theorem' described in Chapter 2 says that the black hole is only characterized by a very few parameters (mass, charge, and angular momentum), and thus we cannot talk about what the black hole is made of.

This lack of knowledge about the nature of what has been sucked in by a black hole may seem like a trivial observation, but it is actually rather profound. Information about a black hole's lunch menu has been fundamentally lost. Any matter which has fallen into the black hole has surrendered its identity. We can't perform measurements on that matter, or discern any details about it.

Black holes and engines

This situation is eerily familiar to those who have studied the beautiful subject of thermodynamics. In that field it is quite common to understand how information can become lost or dissipated through physical processes. Thermodynamics has a long and interesting history. The modern theory began during the industrial revolution when people were trying to work out how to make steam engines more efficient. 'Energy' could be defined in such a way that it was always conserved and could be converted between different forms. This is known as the first law of thermodynamics. However, although you can make some conversions between different types of energy, there are particular conversions you are not permitted to make. For example, although you are allowed to convert mechanical work completely into heat (you do that every time you use the brakes to bring your car to a complete stop), you cannot convert heat completely into mechanical work, which unfortunately is precisely what we would like to do with a steam engine. Therefore a steam engine in a train only succeeds in making a partial conversion of heat from the furnace into mechanical work which turns the wheels. It was ultimately realized that heat is a type of energy involving the random motion of atoms, while mechanical work involves the coordinated motion of some large bit of matter, like a wheel or a piston. Therefore, a crucial component of the nature of heat is randomness: because of the jiggling of atoms in a hot body, you lose track of the motion of the individual atoms. This random motion cannot simply be unrandomized without additional cost. The randomness, or to give it the technical name, *entropy*, in any isolated system never decreases but must always either stay the same or increase in every physical process. (This is the second law of thermodynamics.) One way of looking at this is to say that our information about the world always decreases because we cannot keep track of the motion of all the atoms in a large system. As energy moves from macroscopic scales to microscopic scales, from a simple moving piston to the random motion of huge numbers of atoms, then information is lost to us. Thermodynamics allows us to make this vague-sounding notion completely quantitative. This information loss turns out to be exactly analogous to what we've been describing for matter falling into a black hole.

Although thermodynamics was developed for steam engines, the principles are thought to apply to all processes in the Universe. One of the first people to think about this in connection with black holes was the Oxford physicist Roger Penrose. He reasoned that because a black hole has spin, it might be possible to extract energy from it and thus to use it as some kind of engine. He came up with an ingenious scheme in which matter is thrown towards a spinning black hole in such a way that some of it emerges with more energy than was thrown in. Energy is extracted from the region just outside the event horizon (in fact from the ergosphere discussed in Chapter 3). Penrose's process slows the rotation of the black hole. In principle, an enormous amount of energy can be extracted from a black hole in this way, but of course this is just a thought experiment and so doesn't seem to be at present a practical solution to planet Earth's looming energy crisis! Within a few years of Penrose's work, James Bardeen, Brandon Carter, and Stephen Hawking made a landmark advance and formulated what they called the three laws of black hole dynamics which laid the foundations for Hawking's later thinking on the thermodynamics of black holes, which required the concept of temperature for a black hole which is determined by its mass and spin.

Black holes and entropy

Penrose's insight was a significant impetus and got others thinking about the thermodynamics of black holes. Together with R. M. Floyd, he showed that in his imagined process the area of the black hole's event horizon would tend to increase. Stephen Hawking started working on Penrose's clever scheme. The area depends on the mass and spin (and charge) in a rather complicated way, but Hawking was able to prove that in any physical process this *area* always increases or remains the same. One of the consequences of this intriguing result is that if two black holes coalesce then the area of the black hole event horizon of the merged black holes is larger than the sum of the areas of the two original black hole event horizons. (This is intuitively reassuring because the radius of the event horizon scales with mass, and surface area has a well-known dependence on radius.) This is the same sort of behaviour that we see with entropy in thermodynamics and therefore people began to wonder whether the entropy of a black hole and its area were somehow connected. Is this more than just an interesting analogy? One of John Wheeler's students, Jacob Bekenstein, went ahead and proposed a direct connection in his PhD thesis. Bekenstein used the ideas from the information theory of thermodynamics to argue that the area of a black hole event horizon is proportional to its entropy. (The choice he made means that you take the area of the event horizon and divide by one of physicists' fundamental constants, the Planck area, which is roughly 10^{-70} square metres, and within a numerical factor you get the entropy. This choice of units makes the entropy of a black hole absolutely enormous.)

Initially Hawking didn't believe Bekenstein's results, but on further examination he was able not only to confirm the approach but deepen our understanding of how black hole thermodynamics works. It is perhaps worth understanding how these analyses are done so one can appreciate both their power but also their limitations. The ideal way forward in this field would be to use a combination of quantum mechanics and general relativity, called quantum gravity, to study systems which are both very small, like a singularity in a black hole, but in which gravity plays a big role. Unfortunately we do not have a good theory of quantum gravity at present. A good approach is to use general relativity to model how spacetime curves and then use this together with quantum mechanics to understand the behaviour of particles in the curved spacetime. This was the approach that Hawking took to attempt to understand the thermodynamics of black holes.

Is empty space empty?

The concept of the vacuum (a region where there is 'nothing' there) has had a long and tortuous history. Most of the ancient Greek philosophers hated the idea, on grounds that today seem extraordinarily arcane, but there were a small band of atomists who included the vacuum in their description of the world. Until the scientific renaissance, the idea of the vacuum was therefore very much out of fashion. However, following the invention of the air pump in 1650, the vacuum was something that you could experimentally demonstrate. Even though the amount of air that you could pump out of a vessel in the seventeenth century still gave you a rather poor vacuum by modern standards, the idea of nothingness had become substantially more believable. Once the existence of atoms had been demonstrated beyond all reasonable doubt in the early 20th century, the idea of a region of space with no atoms in it became not only uncontroversial, but inevitable.

No sooner had atoms been demonstrated than a new theory of physics arose: quantum mechanics. One of the surprising consequences of this new theory was that there were fleeting moments when it seems like energy needn't be conserved. The first law of thermodynamics, the grand and seemingly unbreakable principle of physics, insisted that at every moment and at every place there had to be a strict accountancy between energy debits and energy credits. 'Energy must always balance!' thunders the Cosmic Accountant. In fact, it seems that the Universal accountancy rules are more lenient and it is possible to obtain credit. It is perfectly acceptable to borrow energy for a short period of time as long as you pay it back quickly afterwards. The amount you can borrow depends on the duration of the loan, by an amount described by the Heisenberg uncertainty principle. For example, even in the supposedly-empty vacuum it is possible to borrow enough energy to make a particle and anti-particle pair. These two objects can wink into existence and then after an extremely short period annihilate each other, thereby paying the energy back within the maximum allowed time limit (a time interval which is shorter the more energy is borrowed). Such a process goes on everywhere, all the time. It can even be measured! We now understand that the vacuum is actually not empty, but is a soup of these pairs of so-called virtual particles winking in and out of existence. Thus, the vacuum is not sterile and unoccupied, but is teeming with quantum activity.

Black hole evaporation and Hawking radiation

Hawking used the modern theory of the vacuum, quantum field theory, to study its behaviour close to the event horizon of a black hole. His analysis was mathematical but we can picture it in quite a simple way. The essence is that a pair of 'virtual' particles, a particle and its antiparticle (opposite in charge, identical in mass), created close to the event horizon of a black hole may end up becoming torn apart from one another. If one of that pair, either the particle or the anti-particle, falls into the event horizon it will plunge into the singularity and can be never recovered. However, its partner may remain outside the black hole. This particle has lost its virtual partner but it is now nonetheless a real particle and has the possibility of escape. If the particle does escape, rather than falling back in, it forms part of something called *Hawking radiation*. As far as a distant observer is concerned, the black hole has lost mass because a particle has been emitted. What had been realized is that, taking account of quantum field theory, black holes are not completely black, but they can actually emit particles. This argument also applies to photons, and so very weak light (also known as electromagnetic radiation) emerges from a black hole if Hawking's argument is correct.

All bodies at non-zero temperature emit thermal radiation as photons. You do this yourself, which is why you would show up on an infra-red camera even in the dark (and this is why the police and the military use such cameras). The hotter the body, the higher the frequency of the radiation. We emit infra-red radiation, but a red-hot poker is hot enough to emit visible light. Because a black hole emits Hawking radiation, it has a temperature (known as the Hawking temperature) as we have seen earlier, although this is normally incredibly low. A black hole with a mass of one hundred times that of the Sun has a Hawking temperature less than a billionth of a degree above absolute zero (which is 273 degrees below the freezing point of water)! This is one reason why Hawking radiation has not yet been detected: it is incredibly weak. But it is believed to be there.

Hawking radiation does however have an interesting consequence on the evolution of black holes: it is ultimately responsible for a black hole's eventual death. Think again about the two virtual particles. The energy of the real particle which escapes from the black hole has to be positive, but since the virtual particle pair appeared spontaneously from the vacuum, then the virtual particle sucked into the black hole must have negative energy to compensate. Because energy and mass are connected, the net effect of this process is that the black hole has had negative mass added to it, and therefore its mass will have decreased due to the emission of Hawking radiation.

Hawking had therefore discovered a mechanism by which a black hole can evaporate. Slowly, over time, the black hole will emit radiation and lose mass. This process is initially incredibly slow. It turns out that the larger a black hole is the smaller is its '*surface gravity*'. This is because even though the surface gravity depends on mass, which is larger for a big black hole, gravitational attraction follows an inverse-square law and more massive black holes are larger. The net result is that large black holes have very little surface gravity and this equates to a very low temperature. A large black hole therefore emits less Hawking radiation than a small black hole.

However, as a black hole evaporates and loses mass, the amount of Hawking radiation goes up as the surface gravity and hence temperature increases. Assuming the black hole isn't receiving any other energy, this makes the rate of mass loss faster and faster until, at the end of its life, the black hole simply pops out of existence. Thus the life of a black hole ends not with a bang but with that quiet pop. This evaporation process is only possible for black holes whose temperatures are higher than their surroundings. At the current epoch in cosmic history, the temperature of the Universe, measured from the spectral shape of the Cosmic Microwave Background radiation, is 2.7 degrees above absolute zero. Black holes with masses greater than a hundred million million kilos will not evaporate at the current epoch because their temperatures are lower than that of their surroundings. These black holes which have a slender fraction of the mass of the Sun, however, will be able to evaporate when the Universe has cooled more following further expansion. Up to this point in cosmic time, all black holes whose masses were less than one per cent of this slender value would have evaporated away by now.

The black hole information paradox

One question which arises from all of this is what happens to the information stored in the matter that fell into the black hole? One school of thought holds that this information is lost for ever, even if the black hole subsequently evaporates. Another point of view claims that that information is not lost. Because black holes evaporate, the argument goes, the information contained within the original matter that fell into the black hole must somehow be stored in the radiation from the black hole. Thus if you could analyse all the Hawking radiation from a black hole and understand it completely you would be able to reconstruct the details of all the matter that had originally fallen into the black hole. There was a famous bet between, on the one hand Stephen Hawking and Kip Thorne, and John Preskill on the other, about this very matter. Thorne and Hawking took the former position, while Preskill took the latter. The agreement was that the loser would reward the winner with an encyclopaedia of the winner's choice. In 2004, Hawking was sufficiently persuaded by the idea that information could indeed be encoded in the radiation from a black hole that he conceded the bet, supplying Preskill with an encyclopaedia about baseball (whether that constitutes a repository of meaningful information depends on your opinion of baseball); however, the matter is still debated.

Despite all these ingenious theoretical speculations, it is worth saying again that even ordinary Hawking radiation from a black hole has not yet been observed. The history of physics is littered with the relics of old, ingenious but ultimately wrong, theories. Experiments and observation have frequently been surprisingly effective at bringing forth unexpected results. Indeed, observations of spectacular phenomena have emerged that probably no-one at all would have predicted from first principles for black holes. One of the reasons that the faint Hawking radiation has not been observed is that many black holes we know about are at the centres of some of the brightest objects in the Universe, and these black holes are way too massive, and hence way too cold, to evaporate via Hawking radiation. These objects are extraordinarily bright for a completely different reason, which is examined in Chapter 6 and in Chapter 8.

Chapter 6
How do you weigh a black hole?

The Sun, the planets that orbit around it, together with dwarf planets (of which Pluto is the most famous example), asteroids, and comets collectively comprise the Solar System. The Solar System itself orbits within the disc of our Galaxy around its centre of mass at the *Galactic Centre*. The speed at which our Solar System travels around its circular path through the Galactic disc is about 7 km/s, and to complete an entire circuit around the Galactic Centre will take a couple of hundred million years. In addition to this orbital motion, the whole Solar System moves perpendicular to the *Galactic plane*. The kind of motion it exhibits is well known to physicists as simple harmonic motion with the restoring force, which pulls our Solar System back towards the equilibrium position of the plane of the Galaxy, coming from the gravitational pull of the stars and gas that comprise the Galactic disc. At the moment, we are about 45 light-years above this equilibrium point. In about 21 million years from now the Solar System will be at its extreme point 320 light-years above the Galactic plane. 43 million years after that, the Solar System will be back in the mid-plane of the Galaxy. When the Solar System lies in the centre of the Galactic plane then, the Earth will suffer maximum exposure to the cosmic rays that are whizzing around in the plane of the Galaxy, trapped along lines of magnetic field, and travelling around them on some kind of a cross between a helter-skelter and a tramline. There have been speculations that the Sun's motion through the Galactic plane could have been responsible for the mass exinction of dinosaurs. But this kind of speculation is hard to verify or refute because the timescales for this orbital motion are of course rather tricky for human observers, who don't tend to live longer than one century. This is a common problem in observational astronomy when we want to follow some process that changes on timescales much longer than the few centuries over which we've been making astronomical observations of any reasonable accuracy and thoroughness.

There are, however, orbital motions within the Galaxy that are significantly easier to measure, at least in the sense that the relevant timescales are commensurate with the attention spans of humans and their telescopes. Of particular interest in the context of black holes are the orbital motions of the stars in the innermost regions of the Milky Way, that appears in a part of the sky known as Sagittarius A*. Looking into this region, most easily seen from the southern hemisphere, one is looking towards the very centre of our own Galaxy, 27,000 light-years away from us. This is a particularly densely populated region of space, which leads us to two problems when we want to study the Galactic Centre. The first is that there is a relatively high space density of stars and the second is that there is lots of dust.

The first problem means you need to use a measurement technique that enables high resolution imaging, i.e. fine details can be separated from one another in the way that a telephoto lens gives finer detail on a given camera than a wide-angle lens does. Just using a larger telescope is invariably insufficient for this, but there are various techniques developed for untangling the turbulence in Earth's atmosphere through which we inevitably view all celestial objects, unless we put the telescope on a satellite above the atmosphere. Of particular importance is a technique known as *adaptive optics*. This technique corrects for atmospheric variations by observing the blurring of a bright star (called a guide star) and deforming the primary mirror of the telcscope to cancel out this varying blurring. When a bright star isn't available in the part of sky that is of interest, a powerful collimated laser beam can be shone up to excite atoms in the atmosphere and the atmospheric corrections derived from that.

The second issue, the presence of vast quantities of dust towards the Galactic Centre, is problematic because it is hard to see optical light through dust, just as it is hard for ultra-violet light from the Sun to penetrate through the opacity of a sunhat. The solution to this problem is that one needs to observe at infra-red wavelengths rather than visible wavelengths.

How to measure the mass of the black hole at the Galactic Centre

Such infra-red observations have been championed by two groups, one led by Andrea Ghez in California and one led by Reinhard Genzel in Germany. The work of both teams independently provides a wonderfully clear measurement of the mass at the centre of the Galaxy. Figure 14 shows the data from Andrea Ghez and her team. Over the last few years they have made repeated observations right into the very heart of the Galactic Centre and watched how the stars have moved since the last time they observed them. Because the spectral types of these stars are known, their masses are known. Year by year, as the orbital path of each of these stars becomes apparent, the dynamical equations (known as Kepler's laws, the same laws that govern the motion of the planets around our Sun) enable Ghez and her team to solve for each orbit independently and deduce the mass of the 'dark' region that is at the common focus of all these orbits. These independent solutions determine the mass of this dark region rather well. It is now known to be just over 4 million times the mass of our Sun, within a region whose radius is no more than 6 light-hours. Because the object is dark but extraordinarily massive, the only conclusion is that there is a mammoth black hole at the centre of our Galaxy.

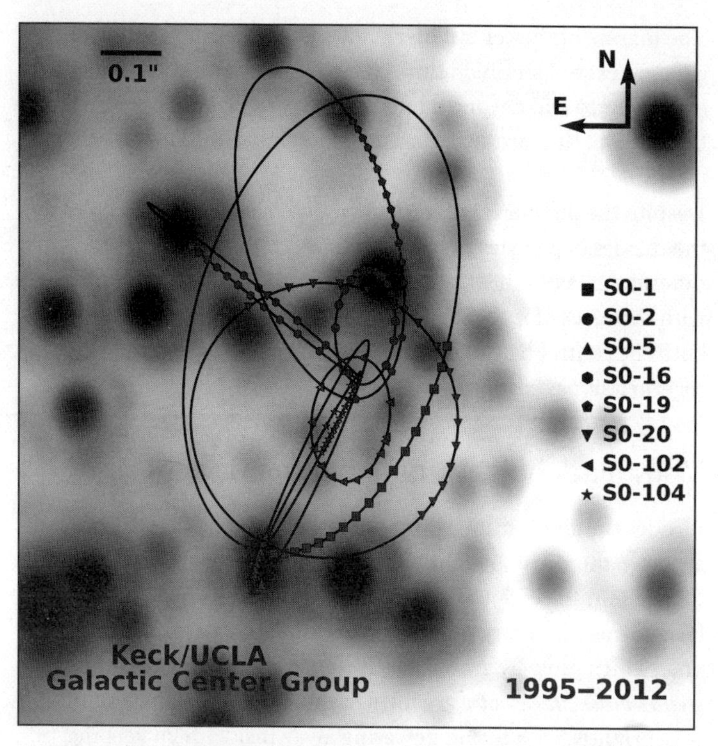

14. Figure showing the successive positions of stars that orbit around the central black hole in our Milky Way.

There is no reason to believe that our Galaxy, the Milky Way, is unique in having a black hole at its centre. On the contrary, it is strongly suspected that all galaxies may well have a black hole at their centres, at least the more massive ones. The reason for this is because of a seemingly fundamental relationship, discovered by John Magorrian, then at the University of Durham, and co-workers, between the mass of a black hole at the centre of a galaxy and the mass of the galaxy itself. Of course the business of measuring the mass of a black hole and the mass of a galaxy is tricky. The technique that works so beautifully at the centre of our Galaxy cannot be applied to external galaxies because they are simply too far away.

The masses of the central black holes at the hearts of elliptical galaxies exceed a million times the mass of our Sun and indeed extend up to and beyond a billion times the mass of our Sun. For this reason, they are often termed *supermassive black holes*.

Despite the difficulties in measuring the masses of black holes and the masses of galaxies, it has been found for a wide range of different galaxies that the mass of the central black hole scales with the mass of its host galaxy. This is thought to suggest that both the central black hole and the galaxy itself grew and evolved together across cosmic time.

Many black holes throughout the Galactic disc

Besides the single, central supermassive black hole at the heart of a galaxy, there are thought to be millions of black holes distributed throughout the extent of each galaxy, and these are believed to have formed in a very different way from the galactic-central ones which grow by gradual accretion of infalling matter. These *stellar mass black holes* are formerly massive stars, once shining very brightly, with fusion powering away inside them keeping them very hot and pressurized, and crucially able to resist gravitational collapse. When their nuclear fuel is all used up, there is no longer any radiation pressure to hold up the star, and nothing to balance the inward force of gravity. For a star with a similar mass as our Sun, the collapse under gravity ultimately results in a compact object known as a white dwarf. The word *compact* has special meaning in astrophysics and connotes that the matter is dense in a way that is utterly distinct from normal matter. By the standards of normal matter, white dwarfs are compact because the matter has been extremely compressed. This matter is ionized, meaning that all the electrons are separate from their parent nuclei, yet cold (normally matter is only ionized at high temperature). The pressure that withstands the persistent inward gravitational pull arises from the electrons refusing to be compressed into too confined a region (a consequence of the

Heisenberg uncertainty principle); the technical name for this effect is 'electron degeneracy pressure'. Had the collapsing star, when it had used up all its fuel, been more massive, then the gravitational infall would have been greater still and the electrons and their counterpart protons would have fused together to form neutrons. These can form a much more compact object than a white dwarf—a neutron star.

But, if we are interested in black holes, then we must turn to stars which are considerably more massive than those which go on to produce white dwarfs or even neutron stars. A star above this mass will be very luminous while its fuel lasts and nuclear fusion can be sustained. Once all the fuel is used up, it's game over for the star and the lights will switch off. The star is now sufficiently massive that the gravitational force can overwhelm even the strong neutron degeneracy pressure and so the collapse is so powerful that even this pressure cannot balance gravity and the collapse leads inexorably to a black hole. The collapse of a massive star is often accompanied by the explosion of a spectacular supernova remnant, leaving a black hole as the only remnant at the original location of the progenitor star. In such explosions many elements, particularly those heavier than iron, are synthesized.

The first black hole to be securely identified from a determination of the masses of the two stars in a binary star system is called V404 Cyg. Jorge Casares and Phil Charles and their co-workers observed the orbits of the two stars very carefully and inferred from their analysis that this binary pair includes a compact object having a mass at least six times greater than the mass of our Sun, and is thus a black hole. (Its mass was later found to be twelve times the mass of the Sun.)

It is possible to make plausible estimates of the numbers of stars in galaxies and their masses. We can then estimate the number of 'stellar-mass' black holes in our Galaxy by considering how many massive stars would have formed early enough in its history to have evolved sufficiently by now to use up all their nuclear fuel via fusion. Even if only a very small proportion of stars in our Galaxy go on to form black holes, with more than 10^{11} objects in the Milky Way that still gives us a lot of black holes.

How can one measure the masses of these black holes that pervade galaxies? In fact for some stellar-remnant black holes, the technique is dynamically very similar to that used for the black hole at the centre of our Galaxy. The reason for this is that a very significant fraction of stars in our Galaxy, and therefore most probably in other galaxies also, come in pairs that formed binary star systems. It is easy to surmise how this might come about: gravitational forces are attractive and many two-body orbits are stable, so once two stars encounter one another and become gravitationally bound together, they are likely to remain so. For a binary system, if we can measure the time taken for the stars to do a complete loop around one another, a time known as the orbital period, and if we know the distance between them, then we are well on the way to finding their masses. If the compact object is in orbit around a normal (fusion fuelled) star of known spectral type and therefore known mass, then the mass of the compact star is straightforward to derive. If a compact object such as a black hole is a singleton and not in a binary, then the lack of dynamical information means that there is no means of inferring its mass and or indeed of determining that it is a black hole. The smallest black hole that we can measure is a few times the mass of our Sun, but the heaviest stellar-mass black holes can exceed a hundred times the mass of our Sun.

The measurement of the mass of a black hole, given modern day technology, is tractable although it still requires a good measure of patience and tenacity. Given that mass is one of essentially only two fundamental physical properties of a black hole such studies get us half-way to characterizing it! However, measuring the spin of a black hole is harder, and in Chapter 7 I describe the heroic efforts that are needed to try and do this.

Chapter 7
Eating more and growing bigger

How fast do they eat?

The popular notion of a black hole 'sucking in everything' from its surroundings is only correct near the event horizon, and even then, only if the angular momentum of the infalling matter isn't too great. Far away from the black hole, the external gravitational field is identical to that of any other spherical body having the same mass. Therefore, a particle can orbit around a black hole in accordance with Newtonian dynamics, just as it would around any other star. What could unravel this pattern of going round and round in circles (or indeed ellipses) and pave the way for more exotic behaviour? The answer is that there is invariably more than one particle orbiting the black hole. The richness of the astrophysical phenomena we observe arises because there is a lot of matter orbiting around a black hole and this matter can interact with itself. What is more, gravity isn't the only law of physics that must be obeyed: so too must the law of conservation of angular momentum. Applying these laws to the bulk quantities of matter that may be attracted towards the black hole gives rise to remarkable observable phenomena, good examples of which are found in the case of exotic objects known as quasars. *Quasars* are objects at the centres of galaxies having a supermassive black hole at their very heart which, because of its effect on nearby matter, can cause it to outshine the collective light from all the stars in one of those galaxies, across all parts of the electromagnetic spectrum. We shall meet quasars, and other examples of 'active galaxies', in Chapter 8, together with scaled-down counterparts of these called microquasars whose black holes are orders of magnitude less massive than those inside quasars. For now let's get back to thinking about the matter around a black hole.

As we have noted, you cannot directly observe an isolated black hole because it simply won't emit light; you can only detect a black hole by its interactions with other material. Any matter falling towards a black hole gains kinetic energy and by *turbulence*, that is to say swirling against other infalling matter doing a similar thing, becomes hot. This heating ionizes the atoms leading to the emission of electromagnetic radiation. Thus, it is the interaction of the black hole on the nearby matter that leads to radiation being emitted from the vicinity of the black hole, rather than direct radiation from the black hole itself.

Black holes are not aloof, non-interacting entities in space. Their gravitational fields attract all matter, whether nearby gas or stars, towards them. Because gravitational attraction increases strongly with proximity, stars are ripped apart if they are unfortunate enough to have a close encounter with a black hole; an example is pictured in Figure 15. A certain fraction of the attracted matter will be entirely swallowed or *accreted* by the black hole. Matter doesn't just accelerate into the black hole whooshing through the event horizon. Rather, there is something of an elaborate courtship ritual as the gravitationally-attracted matter draws near the black hole. Very often it is found that a particular geometry characterizes accreting matter: that of a disc. If the gravitational field were spherically symmetric, the black hole would play no role in determining the plane within which the gas would settle to form an accretion disc—the disc plane would be determined by the nature of the gas flow far from the black hole. If, however, the black hole has spin, accreted gas will eventually settle into the plane perpendicular to its spin axis, regardless of how it flows at

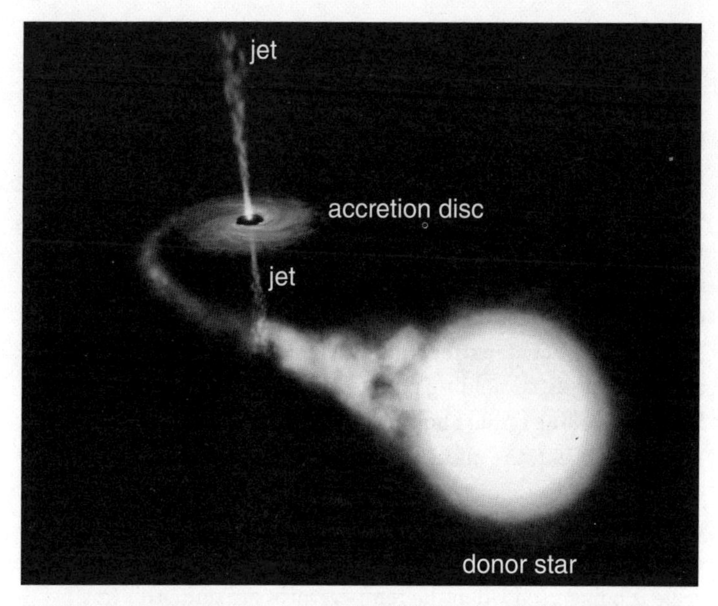

15. Artist's impression of an accretion disc (from which a jet is shown to emanate—see Chapter 8) and a donor star which is being ripped apart by the gravitational tidal forces from the black hole which is at the centre of the accretion disc.

large radii. If there is any rotation at all in the attracted matter, then this must be thought of in terms of the conservation of angular momentum that we met in Chapter 3 when we considered the rotation of material that ultimately collapsed to form a spinning black hole. The rotation means that the matter will be following (fairly circular but actually) spiralling-in orbits as it loses energy. Close to the black hole, the Lense–Thirring effect that we met in Chapter 3 means that at small radii the accretion disc may become aligned with the equatorial plane of the spinning black hole. (In this context, this effect is known as the Bardeen–Petterson effect.)

If gas is a significant component of the collapsing matter then gas atoms can collide with other gas particles on their own orbits and these collisions result in electrons in those atoms being excited to higher energy states. When these electrons fall back to lower energy states they release photons whose energies are precisely the difference between the higher energy level of the electron and the lower energy level it has fallen to. The release of photons means that radiative energy leaves the collapsing gas cloud and so this loses energy. While energy is released in these processes, bulk angular momentum is not. Because angular momentum remains in the system, the coalescing matter continues to rotate in whatever plane conserves the direction of the original net angular momentum. Thus, the attracted matter will invariably form an accretion disc: a rather long-lived holding pattern for material orbiting the black hole. Depending on just how close to the black hole the orbiting material can get, the matter can get so hot that the radiation emitted from the accretion disc actually comprises X-ray photons, corresponding to high temperatures of ten million degrees (it doesn't matter too much whether the Kelvin or Celsius temperature scale is being used when the temperatures are quite this hot!).

A simple analysis of some familiar equations from Newtonian physics shows that the gravitational energy release for a given amount of infalling mass depends on the ratio of its mass multiplied by that of the black hole it is spiralling towards, and how close to the black hole the infalling mass gets. For a given mass of attractor such as a black hole, the closer the infalling mass approaches it, the greater the gravitational potential energy released as can be seen in the cartoon in Figure 16. The energy that is available to be radiated out is the difference between the energy the infalling mass has far away before it is accelerated (calculated using Einstein's famous formula $E = mc^2$, where E is energy, m is mass, and c is the speed of light) and the energy it has at the innermost stable circular orbit of the black hole.

Although fusion holds great hope as a future source of energy for Earth, it can only yield at most 0.7% of the available '$E = mc^2$' energy. In contrast, significantly more of the available rest mass can be released as energy from accreting material, via

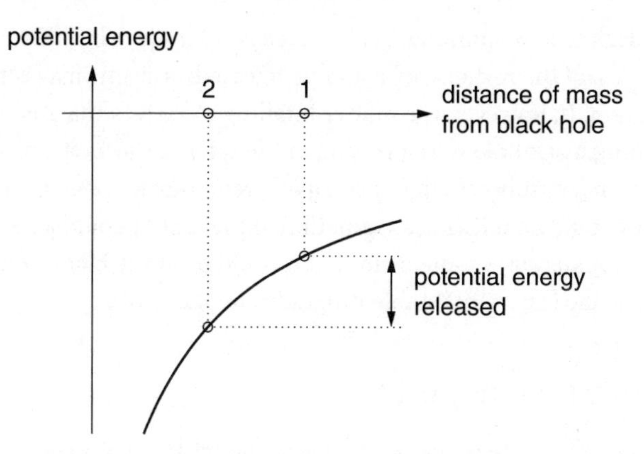

16. Diagram showing how the potential energy of a mass (a test particle) decreases with decreasing distance to a black hole.

electromagnetic or other radiation. Quite how close to a black hole the accreting material can get depends, as described in Chapter 4, on how fast the black hole is spinning. If the black hole is spinning fast, the holding pattern of the material can be orbiting much closer in, on much smaller orbits. In fact, accretion of mass onto a spinning black hole is the most efficient way known of using mass to get energy. This process is thought to be the mechanism by which quasars are fuelled. Quasars are the sites of the most powerful sustained energy release in the Universe and are discussed further in Chapter 8.

I've already mentioned there is an equivalence between mass and energy and for a Schwarzschild (non-rotating) black hole, an amount of energy equivalent to 6% of its original mass could in principle be liberated, and that Roy Kerr's solutions to the Einstein field equations show that the last stable circular orbit has a much smaller radius from the spinning black hole than would a non-rotating black hole of the same mass. In principle, vastly more rotational energy can be extracted from a Kerr black hole, but only if the infalling matter is orbiting in the same sense as the black hole itself. If matter is orbiting in the opposite direction to the way the black hole is spinning, i.e. it is on a retrograde orbit, then not quite 4% of the rest energy could be released as electromagnetic radiation. If, however, the matter infalling towards a maximally spinning black hole were orbiting in the same sense as the black hole were spinning, then in principle a remarkable 42% of the rest energy could be released as radiation, if the matter could lose sufficient angular momentum that it could orbit the black hole as close as the innermost stable prograde circular orbit.

How fast do they eat?

The accretion rate of the black hole at the centre of our Galaxy, in Sagittarius A*, whose discovery we met in Chapter 6, is 100-millionth of the mass of the Sun per year. This doesn't sound very much until you realize that this corresponds to an appetite of 300 Earth masses per year. To account for the typical, immense luminosities of quasars, matter-infall rates amounting to a few times the mass of our Sun each year are required. To account for the typical luminosities of the smaller-scale microquasars that we shall also meet in Chapter 8, the required matter-infall rates might be one millionth of this value.

Another context in which a similar energy extraction process may be taking place is in *gamma-ray bursts*, usually referred to as GRBs. These are sudden flashes of intense beams of gamma rays that seem to be associated with violent explosions in distant galaxies. They were first observed by US satellites in the late 1960s and the received signals were initially suspected to be from Soviet nuclear weapons.

Given the ubiquity of matter spiralling into a black hole via a disc, physicists find it helpful to make simple and instructive calculations to get a handle on the magnitudes of some of the important physical quantities: if one considers a spherical geometry rather than a disc geometry then some interesting limits emerge. A particularly illustrative example comes from the world of stars, which are much better approximations to spheres of plasma than are accretion discs. Sir Arthur Eddington pointed out that the radiation released by the excited electrons colliding with other ions in the hot gas of a star will exert a radiation pressure on any matter that it subsequently intercepts. Photons can 'scatter' (which simply means 'give energy and momentum to') electrons contained in the hot ionized plasma within the interior of a star. This outward pressure is communicated via electrostatic forces (the electrically-charged analogue of the gravitational force) to the positively charged ions such as the nuclei of hydrogen (also known as protons) and the nuclei of helium and other heavier elements that are present.

In the case of a star, the net radiation heads radially outwards and this resulting outward radiation pressure acts oppositely to the gravitational force that pulls matter inward towards the centre. For the more-or-less spherical geometry of a star, there is a maximum limit to the amount of outward radiation pressure before it overwhelms the inward gravitational pull and the star simply blows itself apart. This maximum radiation pressure is known as the *Eddington limit*. Higher radiation pressure inevitably follows from higher luminosity of radiation, and the luminosity of an object can be estimated from its brightness if we know the distance to the object. Therefore, with certain simplifying assumptions including approximating an accretion disc to a sphere, the amount of radiation pressure inside an object can be inferred. This simple method is sometimes used to make an indicative estimate of the mass of the black hole: from the observed luminosity of the radiation to emerge from the surrounding plasma, if it is deemed to be at the maximal limiting value of the 'Eddington luminosity' (above which higher luminosity would give sufficiently high radiation pressure that it would exceed the gravity from the mass within and hence blow itself apart) then the mass can be estimated.

This Eddington luminosity can be thought of in terms of a maximum rate at which matter can accrete, for suitable assumptions about how efficient the process of accretion is. This gives a quantity called the Eddington rate which (for the assumed efficiency) is a maximal value. There are ways of breaking this particular maximum limit, not the least of which is the rejection of the assumption of spherical symmetry (this is fine for a star but manifestly doesn't apply to the disc-geometry of accretion discs that we need to consider in order to understand how black holes grow).

How to measure the speed of rotation within an accretion disc

Because of advances in astronomical technology it is now possible to measure the speed at which material is orbiting a black hole, at least for examples that are relatively close to Earth. One of the big challenges is that it is difficult to obtain information on a sufficiently fine angular scale. The spatial resolution required needs to be at least one hundred, if not one thousand, times finer than that routinely obtained by optical telescopes. In principle, the route to achieving finer resolution with a telescope would be to observe at shorter wavelengths and to build a larger telescope, in particular to reduce the ratio of the wavelength of observation to the diameter of the telescope being used. Unfortunately, the latter gets hideously expensive very quickly while the former takes the usual visible observing wavelengths into the ultra-violet regime, to which the atmosphere of the Earth is rather opaque. The route to achieving a smaller ratio of observing wavelength to telescope diameter is, counter-intuitively, to observe at radio wavelengths (much longer wavelengths than either visible or ultra-violet) for which the atmosphere and ionosphere are usually transparent, and to take the telescope diameter to be most of the Earth's diameter.

There are a few technical issues about this approach which need a little discussion: it turns out that thanks to some very useful mathematics developed by the French mathematician Jean Baptiste Joseph Fourier, it is possible to recover much of the signal that a full telescope aperture would observe, even if the actual

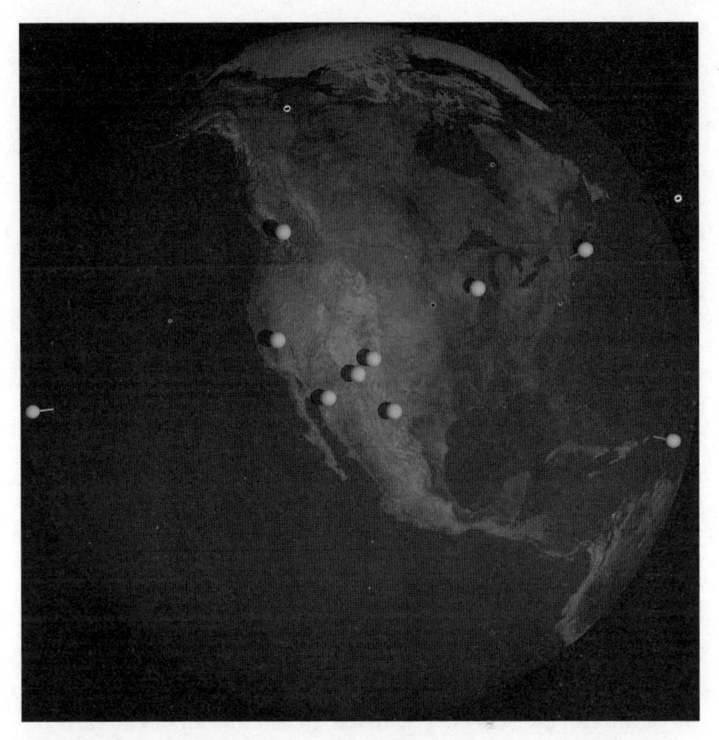

17. Artist's impression of the Very Long Baseline Array (VLBA) of antennas that collectively give images with a resolution equal to that which would be obtained by a telescope with an aperture a significant fraction of the diameter of the Earth.

collecting area only exists in a sparse subset of the full aperture that one would ideally prefer. If the signals from discrete antennas (each looking like an individual telescope—see Figure 17 showing the Very Long Baseline Array, known as the VLBA) are correlated together, it is possible to reconstruct images of small regions of the sky that have detail as fine as that which would be obtained if an Earth-sized telescope could have been fully built. Just to give an idea of how fine this resolution is, suppose that I was standing on top of the Empire State Building in New York, and you were in San Francisco. With this amount of resolution you would be able to resolve detail that is separated by the size of my little finger nail.

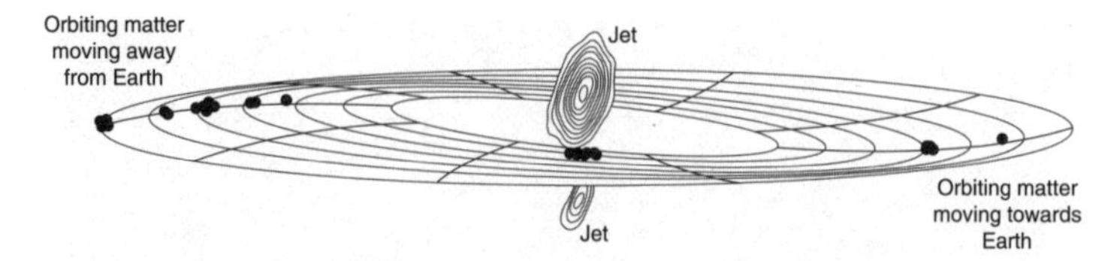

18. The VLBA has measured the distribution of discrete masers orbiting within the accretion disc of the galaxy NGC 4258 (also known as Messier 106) around its central black hole whose mass is 40 million times the mass of our Sun.

(I am glossing over the fact that the Earth is a sphere so there is no direct line of sight between San Francisco and the Empire State Building, but you get the idea.) This means that with instruments like the VLBA we can see individual features less than a light-month apart in other galaxies.

High resolution across an image in a spatial sense, and high resolution in a spectral sense (meaning that one can discern very precisely what the wavelengths of particular features are in a spectrum) is a very powerful combination. Making use of the Doppler effect, a team led by Jim Moran of Harvard University used the VLBA to make observations of the accretion disc surrounding the central black hole of a nearby galaxy known as NGC 4258. They measured the variation in wavelength of a particular spectroscopic signal (a '*water maser*') across the rotating accretion disc and used this redshifting and blueshifting, as the masing matter moved towards and away from the Earth, to detect the variation in the speed with which matter at a given distance orbits around the black hole. These exquisitely beautiful data confirm that the matter orbits around the black hole just as Kepler's laws would describe, and these orbits are depicted in Figure 18.

Swirling matter

In the innermost stable orbit of a black hole whose mass is 100 million times the mass of our Sun, the angular momentum is over 10,000 times smaller than the angular momentum of matter orbiting in a typical galaxy. It is clear that for matter to be accreted by the black hole, this requires the removal of the vast majority of this angular momentum, and this is accomplished by processes within the accretion disc. The orbits in an accretion disc may be regarded as a good approximation to circular although in fact they are subtly and gradually spiralling in. Kepler's laws say that the matter orbiting on the smaller radii will be moving faster than the matter on slightly larger orbits. This differential rotation allows a black hole to absorb the plasma that comprises the accretion disc: the rapidly rotating inner orbits friction burn against the neighbouring material on orbits with slightly larger radii. This difference in velocity will mean that the matter on slightly larger orbits will, by viscous turbulence effects, be dragged along a little faster and so correspondingly the matter on inner orbits will be slightly slowed. Therefore, because orbital motion has increased further out, angular momentum has been transferred to the outer material from the inner material, heating as it does so.

Overall, angular momentum is conserved, and the inner material can systematically lose angular momentum, making it more likely to be swallowed by the black hole. Note that if a blob of orbiting matter has too much angular momentum, it will stay further away from the centre of mass about which it is orbiting: it would be moving too fast to get any closer. What kind of viscous effects might be relevant to the plasma within an accretion disc? Inter-atomic viscosity can be small in this situation—the gaseous plasma of which the accretion disc is comprised is very far removed from the consistency of treacle. In fact, magnetic fields may be very important in transferring angular momentum out of accreting inflow. Where do the magnetic fields come from? The plasma in an accretion disc is very hot, and so the atoms are partially ionized into electrons and positively charged nucleons. Therefore, there are flows of charged particles and moving charges produce magnetic fields, as described by the equations of James Clerk Maxwell. Once even very weak magnetic fields exist, they can be stretched and amplified by differential rotation and modified by the turbulence of the plasma, up to levels at which they can give the required viscosity. This is the basis of what is known as the magnetorotational instability. The importance of this mechanism in this context was first realized by Steve Balbus and John Hawley in the early 1990s when working at the University of Virginia.

By viscous turbulence and probably other means, plasma can eventually lose angular momentum and orbit at smaller radii closer to the black hole. Once the gaseous plasma reaches the innermost stable orbit, no more friction is needed for it to slip down into the black hole, after which it will never be seen again, but it will have augmented the mass and spin of the black hole.

What do accretion discs look like, and how hot are they?

We have seen that viscous and turbulence effects play a significant role in removing angular momentum from the orbiting material so that it can orbit more closely to the black hole and be swallowed by it. A consequence of the viscous action, however, is that the bulk orbital spiralling motion gets converted into random thermal motion and hence the matter heats up. The greater the random thermal motion of matter, the more heat energy it has and the higher its temperature. As mentioned in Chapter 5, wherever there is heat, there will be thermal electromagnetic radiation. Every body emits thermal radiation, unless it is at absolute zero.

Such heating is what is responsible for the highly luminous radiation we observe from accretion discs. For the accretion discs that surround the supermassive black holes that are at the hearts of quasars, the characteristic size of an accretion disc is a billion kilometres and the bulk of the radiation from these accretion discs is in the optical and the ultra-violet region of the spectrum. For the accretion discs that surround the vastly less massive black holes in the so-called microquasars (that are discussed in Chapter 8), the accretion discs are a million times smaller in extent and the radiation is dominated by X-rays. The more massive a black hole is, the larger the innermost stable circular orbit is and hence the cooler the surrounding accretion disc will be.

The maximum temperature in an accretion disc around a supermassive black hole 100 times the mass of our Sun will be around 1 million Kelvin while for a disc around a stellar-mass black hole, it can be up to a factor of 100 higher.

How do you measure how fast a black hole is spinning?

Given you can't actually directly see black holes, you can't see them spinning either. But there are nonetheless two main routes to measuring how fast a black hole is spinning. As discussed in Chapter 4, when black holes spin very fast, it is possible for matter to be in stable orbit around the black hole much closer in than would be possible if they were not spinning. It turns out that matter in these very tight orbits is heated by strong turbulent and viscous effects as it swirls in, and this immense heat can lead to X-rays being emitted, depending on how close to the black hole the matter has swirled in before being swallowed up. General relativity predicts that the shape of the spectral lines is affected by the distance the emitting matter is from the black hole (arising from the gravitational redshift) in a way that has a characteristic signature. This signature arises from fluorescing iron atoms within this matter and the method of extracting information from X-ray light was pioneered by Andrew Fabian of Cambridge University.

These are challenging measurements to interpret, because of many different factors, such as the inclination of the accretion disc with respect to Earth, and indeed the nature of wind and outflowing matter from the surface of the accretion disc, in the vicinity of (along our line of sight to) its inner rim whose characteristics hold the key to unlock information about the black hole that is otherwise inaccessible. Other methods for measuring the spin of stellar mass black holes involve measuring a significant range of the X-ray spectrum and accounting for the different temperatures of the inner regions of the accretion disc (which are hotter) and the regions further out (that become gradually cooler). It is possible to estimate from the shape of the X-ray spectrum the inclination of the disc and from the highest temperature (assuming you know the mass of the black hole, and its distance from Earth) to infer how far from the black hole the innermost material is orbiting. Analogous methods to measure the spin of supermassive black holes at the hearts of quasars are being developed by Christine Done at Durham University. How close that matter is able to orbit (before being swallowed by the black hole) tells you how fast the black hole itself must be spinning.

Black holes are very messy eaters

It transpires that only a fraction (estimated to be 10%, though it can be very significantly higher) of the matter that gets attracted in towards a black hole gets as far as the event horizon and actually gets swallowed. Chapter 8 considers what happens to the matter infalling towards a black hole that doesn't actually get swallowed within the event horizon. From across the accretion disc itself, matter can blow off as a wind; from within the innermost radii of the accretion disc very rapid jets of plasma squirt out at speeds that are really quite close to the speed of light. As Chapter 8 shows, what doesn't get eaten by the black hole gets spun out and spat out rather spectacularly.

Chapter 8
Black holes and spin-offs

Black holes don't just suck

If our eyes could observe the sky at radio or at X-ray wavelengths, we would see that some galaxies are straddled by vast balloons or lobes of plasma. This plasma contains charged particles that move at speeds close to the speed of light and radiate powerfully across a range of wavelengths. The plasma lobes exhibited by some of these galaxies (examples of 'active galaxies') are created by jets, travelling at speeds so fast that they are comparable with the speed of light, that are squirted out from the immediate surroundings of a black hole, outside its event horizon. Roger Penrose showed in general terms how extraction of the spin energy of a black hole from its ergosphere might be possible in principle. Roger Blandford and Roman Znajek have shown explicitly how the energy stored in a spinning black hole could actually be transferred into electric and magnetic fields and thereby provide the power to produce these relativistic jets of plasma. There are also other explanations for the mechanism by which jets are launched from near black holes. However, which of these is correct is the subject of active and exciting current research.

Whatever the mechanism(s) turn out to be, these jets are highly focused, collimated flows ejected from the vicinity of the black hole, but of course outside the event horizon. The regions in between galaxies are not, in fact, empty space. Instead they are filled with a very diffuse and dilute gas termed the *intergalactic medium*. When the jets impinge on the intergalactic medium, shock waves form within which spectacular particle acceleration occurs, and the energized plasma which originated in a jet from near the black hole billows up and flows out of the immediate shock region. As the plasma expands, it imparts enormous quantities of energy to the intergalactic medium. There are many instances of these plasma jets extending over millions of light-years. Thus black holes have tremendous cosmic influence, many light years beyond their event horizons. In this chapter, I will describe the influence and interactions of black holes on and with their surroundings.

As discussed in Chapter 6, at the centre of (probably) most galaxies is a black hole, on to which matter accretes, giving rise to emission of electromagnetic radiation. Such galaxies are called active galaxies. In some of these galaxies, the process of accretion is extremely effective and the resulting emission of radiation extremely luminous. Such galaxies are called quasars (a term which derives from their original identification as 'quasi-stellar radio sources', vastly distant, highly luminous points of radio emission). We now understand that quasars are the sites of the most powerful sustained energy release known in the Universe. Quasars radiate energy across all of the electromagnetic spectrum, from long wavelength radio waves, through optical (visual) wavelengths, to X-rays and beyond. The radio lobes, mentioned above, can be especially dramatic because they extend across distances of over hundreds of thousands of light-years (see Figure 19). The energy radiated at radio wavelengths arises from those large lobes—reservoirs of ultra-hot magnetized plasma, powered by jets that transport energy over vast distances in space. Highly energetic electrons (highly energetic here meaning travelling extremely close to the speed of light) experience forces across their direction of travel from the ambient magnetic fields that pervade the plasma lobes within which they are travelling.

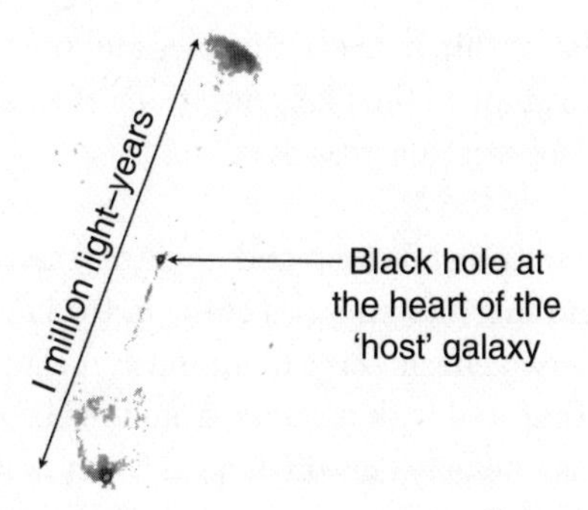

19. This is a radio image of a giant quasar, spanning over one million light-years in extent.

This acceleration causes them to emit photons of radiation (which may be radio, or in rare, highly energetic instances, at shorter wavelengths still, all the way up to X-rays) known as *synchrotron radiation.*

To give a sense of the scale of the power produced by quasars, consider the following values. The LEDs by whose light I am working have a power output of ten watts. They are illuminated by electricity from my local power station which produces a few billion watts (a billion watts is 10^9 watts or a gigawatt). The Sun outputs about 4×10^{26} watts, more than a hundred million billion times that from this power station. Our Galaxy, the Milky Way, contains more than a hundred billion stars, and its power output is approaching 10^{37} watts. But the power produced by a quasar can exceed even the Galactic power output by more than a factor of 100. Remember, this power is being emitted not by a galaxy of one hundred billion stars but by the processes going on around a single black hole. Such radiation could do considerable damage to the health of living creatures here on Earth, so it is just as well for us that there are no examples of such powerful quasars too near our Galaxy!

Jets in quasars are thought to persist for a billion years or less, an idea that comes from estimates of the speed at which these objects' jets grow and from measurements of the size they have grown out to. A simple relationship between distance and time and speed therefore gives a guide to the likely durations of jet activity in the quasars that are observed across the cosmos.

As these radio-emitting lobes expand, their magnetic fields weaken as do the 'internal' energies of the individual electrons in the lobes. These two effects serve to diminish the intensity of the radiation with time and with distance from the black hole; how dramatically this intensity falls off depends on how many highly energized electrons there are compared with how many less energetic ones there are. It's a property of synchrotron radiation that the lower the magnetic field strength is, the more energetic the electrons need to be to produce the radiation at the wavelength that your radio telescope is tuned to receive at. This compounds the diminishing of the synchrotron radiation as the plasma lobes expand into outer space. Not only do the electrons lose energy as the plasma expands, but because the magnetic field strength is weakening, only increasingly energetic electrons are relevant to what is observed by your telescope and, very often, there are vastly fewer of these than there are of the lower energy electrons anyway. As far as radio lobes of quasars are concerned, the lights can go out really quite rapidly.

The show isn't over, but the spectacle does move over to a different waveband. Something rather remarkable happens: the lobes light up in X-rays. This happens via a scattering process known as *inverse Compton scattering*. In the presence of a sufficiently large magnetic field, electrons can emit synchrotron radiation and thereby lose energy. Another mechanism of losing energy that is relevant to our discussion here happens via the interaction of these electrons with photons that comprise the Cosmic Microwave Background (CMB), the radiation that is left over from the Big Bang and which currently bathes the Universe in a cool microwave glow. It is possible for such an electron to collide with a photon from the CMB so that the photon ends up with a lot more energy than it had before the collision and the electron ends up with a lot less energy than it had before the collision (energy is conserved overall, remember). Of particular interest is that when the energies of the rapidly moving electrons reduce to a mere one thousand times the energy of an electron at rest (having previously been a hundred or a thousand times higher than this) their energies are perfectly matched so that they will upscatter CMB photons into the X-ray photons. The interaction of an energetic electron with a low-energy photon to yield a high-energy photon is somewhat analogous to the situation in snooker where the white cue ball (imagine this is an electron) collides with one of the red snooker balls (for the purposes of this illustration please overlook the fact that this ball isn't moving at the speed of light!) and the red ball gains a lot of energy at the expense of the cue ball. Whereas (hopefully) the red ball ends up in one of the pockets on the snooker table, the photon (which originally had a wavelength of about a millimetre) acquires about one million times as much energy as it had before the collision so that its wavelength becomes a million times shorter.

The *Chandra* satellite, launched by NASA in 1999, is sensitive to X-ray wavelengths and in fact can detect pairs of dumb-bell lobes in the X-rays just as a radio telescope can detect these double structures at cm-wavelengths. Figures 20 and 21 show in contour form double structures observed at radio wavelengths and in greyscale form the double structures in X-rays.

In fact if we were able to monitor the life cycle of one of these quasars throughout all these evolutionary stages (analogously to how a biologist might observe the life cycle of the frog from frogspawn, to tadpoles, to tadpoles with little legs, to little frogs with stumpy tails, to larger frogs to dead frogs) we would observe a cross-over from the double structures being radiant at radio wavelengths to becoming increasingly dominant in the X-ray region. First the radio structures would fade beyond detectability then the X-ray structures would fade beyond detectability. Of course, if the jets were to re-start, for example if the black hole

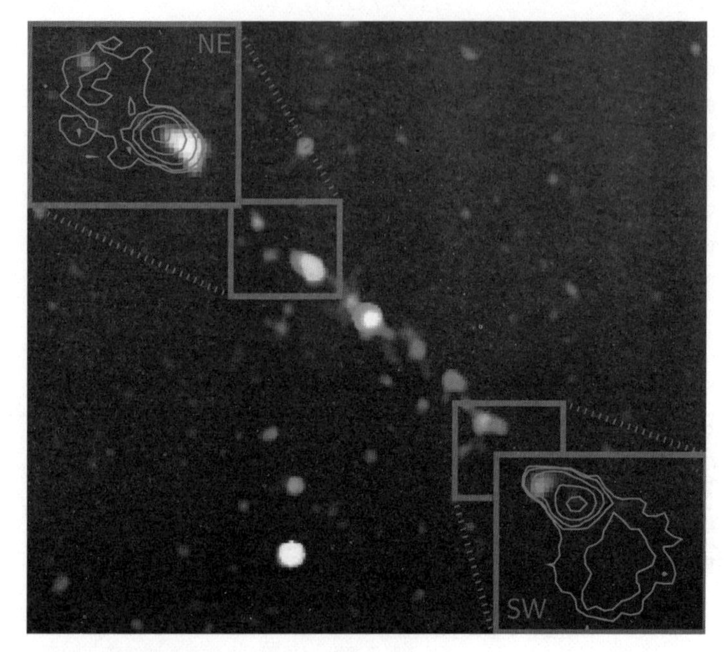

20. This giant quasar is half a million light-years in extent, and has a double-lobe structure at both radio (shown as contour lines) and X-ray (shown in greyscale) wavelengths.

were to get more fuel, then the jets would fuel new radio-emitting double lobes and then X-ray emitting lobes again. As we have seen in Figures 20 and 21, in some quasars we can see both the radio and the X-ray double structures at the same time but in others, only one or the other (Figure 22). In a couple of remarkable cases we see the X-ray double structure corresponding to a previous incarnation of jet activity, but also some new radio activity, at a different angle because the direction along which the oppositely-directed jets are launched has swung round, i.e. it has precessed; an example of this phenomenon is seen in Figure 21.

The steadiness of the jet axis of many quasars and radio galaxies is a pointer to the steadiness of the spin of the supermassive black hole, acting like a gyroscope. Why some of these jet axes should

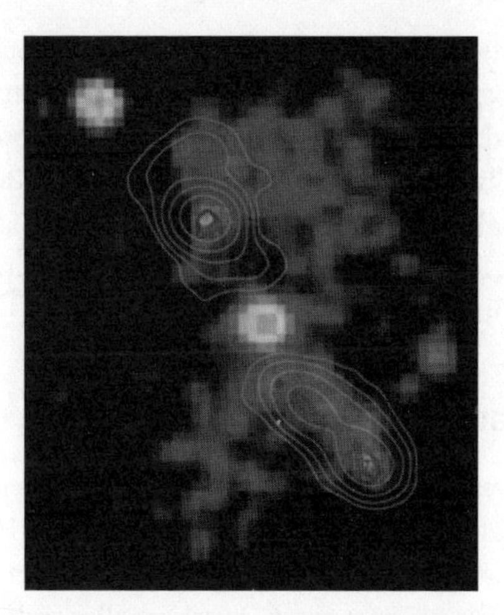

21. The double-lobe structure observed in this quasar at radio wavelengths [contours] showing the more recent activity to be differently oriented from that showing at X-ray energies [greyscale] (the relic emission revealed by inverse Compton scattering of CMB photons) suggesting that the jet axis may have precessed as the jet axes in microquasars do.

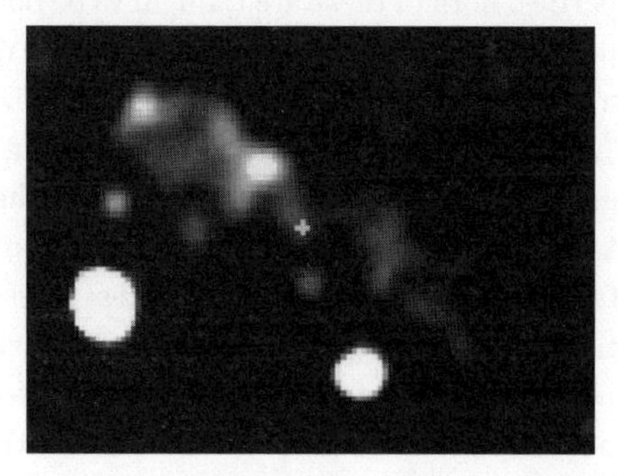

22. This is an X-ray image and shows the double-lobe structure straddling this galaxy which is only detectable at X-ray wavelengths.

precess but not others will be answered when we can discover what controls the angular momentum of the jets at the launch point near the black hole. Whether this is the spin axis of the black hole itself, or whether it is the angular momentum vector of the inner part of the accretion disc, compounded no doubt by the Lense–Thirring or Bardeen–Petterson effects I mentioned in Chapters 3 and 7 respectively, is not yet clear and more data are required to fully elucidate the observed behaviour. But, there are clues from smaller objects closer to home that may suggest that the precession of jet axes is everything to do with the accretion disc's angular momentum.

Microquasars

The quasars we have been discussing so far are all supermassive black holes that lie at the centres of active galaxies. However, it turns out that there is another class of objects that behave very similarly but are on a much, much smaller scale. These lower mass black holes can be observed rather closer to home, indeed located within our own Milky Way Galaxy, and they are called 'microquasars'. Although the difference in scale size is vast, microquasars in our Galaxy and extragalactic quasars at the centres of other galaxies are both sources of plasma jets with analogous physical properties. Both of these are thought to be powered by the gravitational infall of matter onto a black hole. In the case of a microquasar, the black hole has a mass comparable with that of the Sun. In the case of a powerful extragalactic quasar, the mass of its black hole can be a hundred million times larger than the mass of our Sun. As far as the astrophysicist is concerned, an important advantage of the local examples is that being less massive, they evolve much more rapidly, on timescales of days rather than millions of years in the case of quasars. Nonetheless, as in the case of quasars, the jets which are squirted out from near the centre of all the activity are launched from outside the event horizon, and very likely from the innermost edge of the accretion disc.

Complex mechanisms are at play, and there isn't a simple relationship between the speed at which a jet is launched and the mass of the black hole with which it is associated. In the course of monitoring the jets in the black hole microquasar called Cygnus

X-3 there are occasions when the speeds at which the jet plasma moves away from the black hole are found to vary. This has been measured by time-lapse astronomical measurements in which observations at successive times allow us to determine how fast the jet plasma is hurtling away from the vicinity of the black hole. Such measurements have shown on one occasion the jet speed to be 81% of the speed of light whereas four years later to be 67% of that speed. There is no suggestion that the jet speed is merely reducing with time, since fast and slower jets speeds in this microquasar appear to have been witnessed on a number of occasions since its discovery. Varying jet speeds seem to characterize another well known microquasar in our Galaxy, called SS433, that I shall describe in more detail below. The jet speed in this microquasar seems to change quite a bit as well, indeed it can be anywhere between 20 and 30% of the speed of light over just a few days.

The beauty of symmetry

Figure 23 shows a radio image of SS433, a microquasar in the Galaxy, which is a mere 18,000 light-years distant from us. The striking zigzag/corkscrew pattern is the structure of the plasma jets as they appear to us on the plane of the sky. The individual bolides of plasma that make up the jets are moving at tremendous speeds that vary between 20 and 30% of the speed of light. The directions along which the bolides are moving varies with time in a very persistently periodic way. In fact the axis along which the jets are launched precesses in much the same way as does the paddle of a kayakist, in the frame of reference of the kayak, except on a timescale of six months rather than several seconds. This same behaviour is apparently taking place in at least some quasars (see Figure 21) albeit in that case in such slow motion that we are unable to appropriately time-sample the changes taking place.

The detailed appearance of the zigzag/corkscrew pattern on the sky depends directly on the physical motions of the bolides, as well

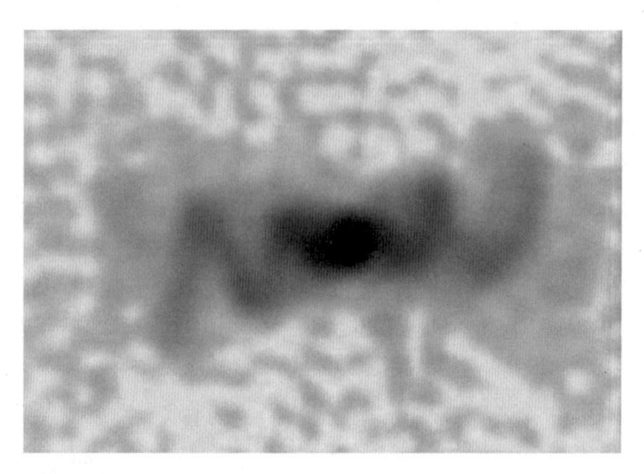

23. The jets of the microquasar SS433 as they appear at radio wavelengths.

as the time when the observation is made. One of the remarkable features of the jets is their symmetry: the physical motions of the components in the eastern jet are equal and opposite to those in the western jet: when one bolide of plasma is at 28% of the speed of light, so too is its counterpart in the oppositely-directed jet; for a different bolide of plasma moving at 22% of the speed of light, so too will its counterpart in the oppositely-directed jet. The fact that one jet appears to have a zigzag structure while the other appears to have a rather different corkscrew pattern is a consequence of the jet plasma always moving at speeds comparable with the speed of light, and well-known relativistic aberrations that occur under such circumstances. The power radiated by this microquasar is rather modest relative to that of an extragalactic quasar but it is still vast in comparison to the power of the Sun which seems somewhat puny, having a total luminosity of only 4×10^{26} watts, a factor of a hundred thousand smaller than that radiated from the microquasar in Figure 23.

Jet launch

The Virgo Cluster is a cluster of well over a thousand galaxies just over fifty million light-years distant from the Milky Way. At its

24. A jet of plasma squirted out at speeds close to that of light, from the supermassive black hole at the heart of the M87 galaxy.

heart is a giant galaxy called M87 (an abbreviation of Messier 87, listed in the catalogue produced by the French astronomer Charles Messier). And, at its heart, is a supermassive black hole whose mass is three billion times that of our Sun. Emanating from this is a strong straight jet, as shown in Figure 24.

This jet is readily visible at optical wavelengths, at radio wavelengths, and at X-ray wavelengths. It is thought that the infalling matter accretes at a rate of two to three Sun's worth of mass per year, onto the very central nucleus where an accretion disc of the sort described in Chapter 6 is thought to be at work. The speed at which this jet propagates away from its launch point, likely at the innermost region of the accretion disc, is very close to the speed of light, and so we refer to it as a relativistic jet. Jet speeds close to the speed of light are revealed by successive monitoring with the VLBA instrument that I introduced in Chapter 7, and the Hubble Space Telescope and Chandra X-ray satellites which are each above Earth's atmosphere and thus attain higher sensitivity than if they were on the ground. At 50 million light-years from Earth an object moving at the speed of light would move across the sky at four milli-arcseconds per year. When we consider that an arcsecond is 1/3600 of a degree, then four-thousandths of this may sound like a tiny angle to measure, but such separations are easily resolvable with an instrument like the VLBA. The VLBA has already imaged the base of this jet to within less than about thirty Schwarzschild radii of its supermassive black hole.

Figure 25 shows an example of the lobes and plumes of radio emitting plasma fed by the relativistic jets from the supermassive black hole in M87.

By way of further illustration that expansive lobes are associated with relativistic jets, Figure 26 shows an example that extends 6 degrees across the sky, and is shown to give a sense of scale with respect to the telescope array used to make the observation. The telescope, used by Ilana Feain and her colleagues, was the Australia Telescope Compact Array.

The mechanisms by which relativistic jets are launched from the vicinity of a black hole remain much closer to conjecture than to acceptance beyond all reasonable doubt. Nonetheless, various independent lines of research by entirely independent teams based in different countries around the world seem to be implying that the preponderance of evidence is that the basic emerging details are correct. Beyond the broad picture, however, the mechanisms and their detailed functioning are conjectural, but being patiently tested amid insufficient photons and selection effects. Proof doesn't belong in science but evidence very much does. We are hindered because even the most advanced imaging techniques deployed today cannot separate and resolve the smallest regions where most of the energy is released, but this is where numerical simulations on powerful computers can transcend the limitations of current technology. Indeed results from simulations of jet launch from accretion discs that fully account for general relativity effects are just being published. These simulations, with known input ingredients and axioms, allow jets and discs to evolve to size-scales where their properties can be confronted against state-of-the-art observations.

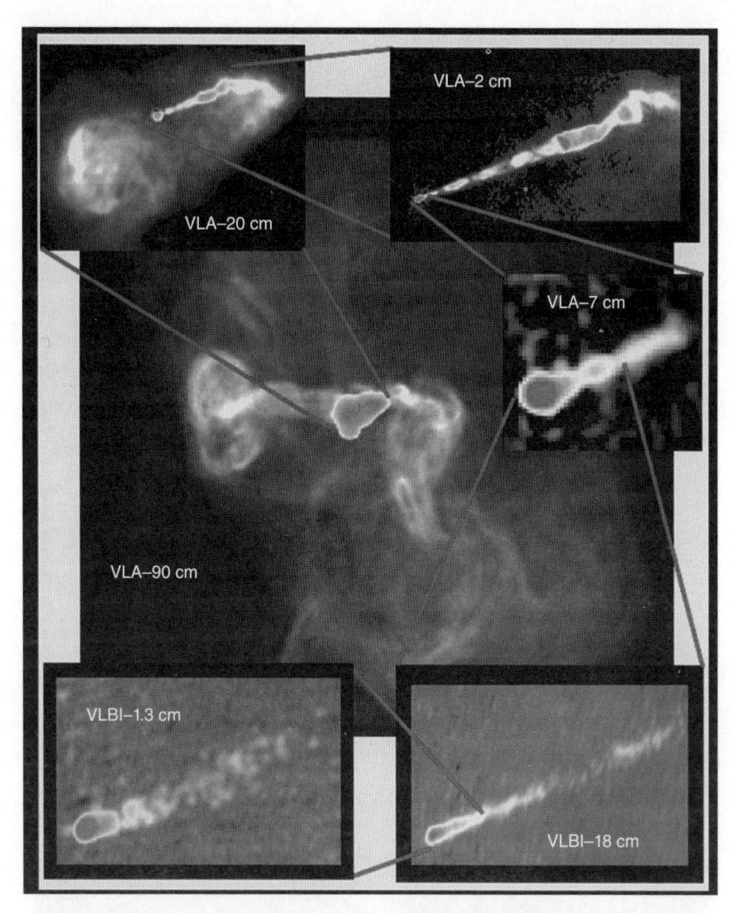

25. The radio-emitting lobes that and plumes are fed by the relativistic jet emanating out of the supermassive black hole at the centre of the M87 galaxy.

So what do we now know about the masses of black holes in the Universe? It seems that they fall into two main classes. First, those that have masses similar to those of stars. These stellar mass black holes come in between around three to thirty times the mass of our Sun and come from stars that have burned all their fuel.

26. Composite picture showing an optical image of the moon and the Australia Telescope Compact Array, and a radio image of Centaurus A.

Then there are the supermassive black holes which go all the way up to about ten billion solar masses. As we have discussed, these are found in the centres of galaxies including our own and are responsible for the extraordinary phenomena of active galaxies and quasars.

We have talked about things falling into a black hole, but what happens when a black hole falls into a black hole? This is not an abstract question, since it is known that black hole binaries can exist. In such objects two black holes are in orbit around each other. It is thought that, because of the emission of gravitational radiation, the black holes in a binary will begin to lose energy and spiral into each other. In the final stages of this spiralling, general relativity is pushed to breaking point and the black holes suddenly coalesce into a single black hole with a common event horizon. The energy released in the merger of two supermassive black holes in a binary system is staggering, potentially more than all the light in all the stars in the visible Universe. Most of this energy is dumped into gravitational waves, ripples in the curvature of spacetime, which propagate across the Universe at the speed of light. The hunt is on for evidence of these waves. The idea is that as a gravitational wave passes by a material object, like a long rod, its length will fluctuate up and down as the ripples in spacetime curvature flow through it. If you can measure these tiny length changes, using a technique such as laser interferometry, then you have got a method to detect gravitational waves produced elsewhere in the Universe. Both ground- and space-based gravitational wave detectors, examples of which have been built and more of which are planned, have the potential to pick up signals from black hole mergers. In fact, gravitational waves are so difficult to detect that you need a very strong source to have any chance of such experiments working, and a black hole merger is high on the list of candidates for such strong sources. At the time of writing, gravitational waves have not yet been directly detected, but the experiments are ongoing.

Our best theory of gravity, which comes from Einstein's general theory of relativity, has survived countless tests since its discovery in 1915. It has been shown to give far better agreement to experiment than Newton's theory which it supplanted. However, if general relativity is ever going to be tested up to its limits, you can confidently expect that black holes will prove to be the ultimate testing ground of this cornerstone of modern physics. Where gravity is the most intense in the smallest region of space, so that quantum effects should be important, is exactly where general relativity might break down. However, it might also be that general relativity breaks down on large scales in the Universe. Of course, a hot topic at present is the completeness of general relativity to explain the accelerated expansion of the Universe on the largest scales. Possible deviations away from general relativity are being discussed in connection with accelerated expansion and dark energy. If gravitational waves are detected from the mergers of black holes, or if observations extend our understanding of the fundamental physics which occurs in the vicinity of these fascinating objects, then there's a good chance that we will be able to see how well Einstein's theory holds up or whether it needs to be replaced by something new.

Why do we study black holes?

There are a number of reasons for investigating black holes and one is that they open up the exploration of physics parameter space that is otherwise simply inaccessble to the budgets of even an international consortium. Black hole systems represent the most extreme environments that we can explore, and as such probe the extremes of physics. They bring together both general relativity and quantum physics whose unification has not yet been achieved and remains very much a frontier of physics. A second reason is that trying to understand black hole phenomena arouses fascination in scientists and many thoughtful lay people, and provides a route by which many people can be stimulated by science and motivated to learn about the almighty magnificence of the Universe around us. A third and perhaps surprising reason is Earthly spin-offs. How could black hole research conceivably change our lives? The answer is that it has already done so. As I type these final sentences of this little book into my laptop, it simultaneously backs up my work onto my University server via the 802.11 WiFi protocol. This intricate and clever technology emerged directly out of a search for a particular signature, at radio wavelengths, of exploding black holes led by Ron Ekers to test a model suggested by (now Astronomer Royal, Lord) Martin Rees. Ingenious radio engineers in Australia, led by John O'Sullivan, in the course of devising interference suppression algorithms for the tricky business of detecting subtle signals from distant space realized that these could be applied to transform communication here on Earth. Black holes therefore have the power to rewrite physics, reinvigorate our imagination and even revolutionize our technology. There are many spin-offs from black holes—way beyond their event horizons.

Further reading

M. Begelman and M. Rees, *Gravity's Fatal Attraction*, 2nd ed. (Cambridge University Press, 2010).

J. Binney, *Astrophysics: A Very Short Introduction* (Oxford University Press, 2015).

J. B. Hartle, *Gravity* (Addison Wesley, 2003).

A. King, *Stars: A Very Short Introduction* (Oxford University Press, 2012).

A. Liddle, *An Introduction to Modern Cosmology*, 3rd edn. (Wiley-Blackwell, 2015).

F. Melia, *The Galactic Supermassive Black Hole* (Princeton University Press, 2007).

D. Raine and E. Thomas, *Black Holes: An Introduction*, 2nd edn. (Imperial College Press, 2010).

C. Scarf, *Gravity's Engines: How Bubble-Blowing Black Holes Rule Galaxies, Stars, and Life in the Cosmos* (Scientific American/ Farrar, Straus and Giroux; Reprint edition, 2013).

R. Stannard, *Relativity: A Very Short Introduction* (Oxford University Press, 2008).

A. Steane, *The Wonderful World of Relativity: A Precise Guide for the General Reader* (Oxford University Press, 2011).

K. Thorne, *Black Holes and Time Warps* (W. W. Norton, 1994).

Index

V

W

Z

激发个人成长

多年以来，千千万万有经验的读者，都会定期查看熊猫君家的最新书目，挑选满足自己成长需求的新书。

读客图书以“激发个人成长”为使命，在以下三个方面为您精选优质图书：

1. 精神成长

熊猫君家精彩绝伦的小说文库和人文类图书，帮助你成为永远充满梦想、勇气和爱的人！

2. 知识结构成长

熊猫君家的历史类、社科类图书，帮助你了解从宇宙诞生、文明演变直至今日世界之形成的方方面面。

3. 工作技能成长

熊猫君家的经管类、家教类图书，指引你更好地工作、更有效率地生活，减少人生中的烦恼。

每一本读客图书都轻松好读，精彩绝伦，充满无穷阅读乐趣！

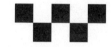